STUDIES IN PHILOSOPHY AND IN THE HISTORY OF SCIENCE:

Essays In Honor Of Max Fisch

Edited by Richard Tursman
With a Preface by D. W. Gotshalk

Coronado Press 1970

by Coronado Press
Box 32
Lawrence, Kansas 66044
USA

Standard Book Number 87291–007–5

Manufactured in the USA

ὅπως εἰς δύναμιν ἡ περὶ τὰ
ἀνθρώπινα φιλοσοφία τελειωθῇ.

CONTENTS

Preface

by D. W. Gotshalk

The fourteen essays in this volume were written by philosophers who were formerly graduate students of Max Fisch at the University of Illinois. The most obvious characteristic of the essays is their great variety. Their subjects range from Aristotle, Harvey, Hegel, Agassiz, and Peirce, to Freedom, The Justification of Morality, A Third Century Taoist Treatise,Synaesthesia, and the Frequency Theory of Probability. In this variety they represent not only the diverse predilections of their authors but also the wide scope of the scholarly interests of the philosopher they honor.

Born in 1900 at Elma, Washington, the son of a lumberjack, Max Fisch received his high school training in San Francisco, and his undergraduate education at Butler University in Indianapolis. He then entered Cornell where he took his doctor of philosophy degree (1930). Fisch came to Ithaca with a knowledge of Greek, Latin, German, French and Spanish, an omnivorous interest in books old and new, a critical awareness of Western philosophy acquired at Butler under Elijah Jordan, and the ambition to prepare himself to be a helpful teacher of philosophy. At Cornell his graduate work was mainly in the history of philosophy with concentration on the ancient Greek and Roman period, and his doctoral dissertation was on the influence of Stoicism on Roman Law.

After Cornell came teaching at Western Reserve University (1928-1943), war years as curator of rare books and chief of the History of Medicine Division of the U.S. Army Medical Library which had been moved to Cleveland (1942-1946), then the University of Illinois (1946 on). At Western Reserve a full teaching schedule presented the opportunity to develop the strong interest in American and modern philosophy aroused by Jordan, and under the stimulus of his colleague Thomas Bergin, Fisch discovered the possibility of using his Graeco-Roman learning to illumine the thought of certain Italian thinkers such as Vico. These two large areas, American philosophy and Italian thought from the Renaissance onward, were to remain the major fields of his scholarly interest, although his publications were also to include studies in ancient Greek and Roman philosophy, several articles on Vesalius and a book on Nicolaus Pol, articles and reviews on various aspects of the history of science, and papers on certain perennial philosophical problems.

Without doubt Jordan's teachings and personality have been the major philosophical influence on Fisch, although his wife, Ruth, a Latin scholar, has been a tireless helper in all his work. Jordan's strong respect for the major figures of of Western traditional philosophy, particularly Plato and Kant, his corporative or institutional theory of human existence and action, his sharp figures of Western traditional philosophy, particularly Plato and Kant, his corporative or institutional theory of human existence and action, his sharp criticism of contemporary philosophy and institutions, and his dedication to philosophy as vital to American life, deeply impressed Fisch, as did Jordan himself, an ardent philosopher isolated and neglected in the mid-west American wilderness.

There were however important dissimilarities between the two men. Jordan was a man of extremes. He could purr like a kitten and among admirers be the soul of sweetness. But he had the claws of a tiger, and in his writings would tear to pieces with ferocity and contempt all thought he believed to be specious and all ways of life including American business life that he believed to be harmful. Deep down in his pupil perhaps the same combination of antithetical qualities might be found. But a quiet friendly gentleness and reserve are more prominent, and only an occasional sardonic remark or an unexpected advocacy of a strong dissent indicates the cutting critical mind working underneath.

Two other dissimilarities are important, paralleling the first. Jordan's chief labor was to expound his own philosophy, Fisch's to describe the life and thought of others, saving his own philosophy for special occasional papers. Furthermore, the form each used in his exposition was as different as the content. Jordan set forth his philosophy in a prose style generally attributed to German philosophers, using long convoluted sentences, as opaque at times as the more mystical passages of the symphonic Bruckner. In contrast, Fisch developed a polished prose style distinguished by clarity, meticulous accuracy, and economy of expression. Everything is stated simply and precisely, and everything that need be stated is there. Fisch's prose is chamber music. Its texture exhibits the niceties of intricately related strings. The turgid proclamatory assertiveness of symphonic brass and percussion is absent. But lucidity and masterly use of details give full measure to what is being said.

Today, Max Fisch is one of two or three American philosophers who stand out as exceptionally fine historical scholars. His range is probably wider than any, his scrupulousness and cogency the equal of the best. His works on Vico, Peirce, American classical philosophy, and the history of medical science in the Renaissance are permanent additions to our understanding of these subjects. His papers on philosophy, philosophy of history, and philosophy of law state a philosophical position that, inspired by Jordan, is a clearer if briefer challenge to the individualistic and divisive modes of action and thought that Jordan challenged in his day. A student possessing the finest scholarly talent and equipment, Fisch has also been a thinker aware of the deficiencies of current philosophical and social movements, and ready with suggestive alternatives. In addition, he has achieved his initial modest goal set at Cornell of being a helpful teacher of philosophy, as the essays in this volume attest.

D. W. Gotshalk,
Professor Emeritus,
University of Illinois,
Urbana, Illinois.

Editorial Note

Max Fisch is an excellent human being, a brilliant scholar, and probably the best philosopher in America today. There are those who, while willing to admit that Fisch is a wonderful person and a fine scholar, think that he is merely a good philosopher and not the best. I think that such critics have failed to notice the subtlety with which Fisch writes, the rich variety of philosophically important elements which he incorporates into his writings, and, indeed, the fact that in order to have a philosophy of high quality it is necessary to take into account such a wide range of elements — some of which are profound. So that the reader may become better acquainted with Fisch's life, work and thought, we have included in this volume two articles which are representative of his philosophical thought, a bio-bibliographical sketch, a bibliography of his writings, and a list of theses directed by him at the University of Illinois.

Festschriften are difficult to get published these days: we are indebted to the understanding and generosity of Mr. John E. Longhurst and the Coronado Press for making this publication possible.

R.T.

Four Uses of "Freedom"

by Robert A. Cornett

In this paper I shall discuss briefly, or in outline, four common uses of "freedom" and its corresponding adjective, "free". I shall illustrate these uses by reference to four of my own freedoms, or acts relative to which one could properly say that I am free. The four freedoms are my freedom to play a sitar, my freedom to smoke "grass" (marihuana), my freedom to eat grass (real grass) and my freedom to play a guitar. The following facts are essential to the discussion. I cannot play a sitar, it is illegal to have marihuana in my possession and therefore to smoke it, I have no desire to eat grass, and I can play a guitar after a fashion and enjoy playing it in that fashion. It will be useful also to note that playing the sitar, eating grass and playing the guitar are legal and that because of this, and for other reasons, to do so would not result in the imposition of penalties upon the performer by his fellowmen.

If a friend, knowing that I cannot play a sitar, should say to me, "You are, nevertheless, free to play a sitar," I should take him to mean that there is nothing in my cultural circumstances to hinder me from playing it. And I might also infer, from his mentioning it, that possibly there was something in his cultural circumstances which hindered him from playing the sitar. Perhaps his wife was opposed to it and had threatened to leave him if he took it up, or his neighbor detested sitar playing and had threatened to sue him if he should attempt to play, or his organizational affiliations prevented it, barring sitar players from membership, or possibly there was a legal ban on sitar playing where he lived, with a fine or imprisonment or both for those who were apprehended. At any rate I should take him to mean that there was no condition similar to these in my own circumstances which would hinder or prevent my playing the sitar. And I should judge in any case that this would be anyone's meaning when saying that I was "free" to do something which was for me an impossible act. And, of course, for me and for anyone there are innumerable such acts which are permitted for us by our cultural circumstances, the performance of which involves no threat or likelihood of penalty from one source or another, yet which are for us impossible, due either to a lack of ability or to the non-existence of the means necessary for the performance of the act. I may refer to freedom in this sense, that is, as consisting of the acts which are permitted for me by my cultural circumstances and including those which are not possible for me, as my *formal* freedom. Thus I am free to play the sitar in this formal sense, although it is not possible for me to do so.

However, it would not be unusual for someone to suggest to me that I am also "free" to smoke "grass," or, in other words, "free" to do something which is not permitted by my cultural circumstances. In this particular case the act is not only illegal and subject to penalty by law, but also subject to penalty from other sources. It is a possible threat to my institutional relations, to my social status, to my income, and perhaps even to established friendships. Still, someone might argue that I am "free" to do this in the same sense that I am "free" to participate in illegal civil rights activities, "free" to run the stop sign when no one is looking, "free" to cheat on my income tax report, "free" to tell lies, break promises, assassinate the president, or to carry out innumerable illegal or immoral, or unconventional acts subject to some kind of penalty imposed by others, either officially or unofficially. However, when someone uses "freedom" in this sense, to suggest to me that I am "free"to carry out an unpermitted act, his only purpose or meaning could be to call attention to the fact that the act is possible for me. In other words, "freedom" used in this way means, essentially, "can," or, that at that particular moment nothing is hindering or preventing me from carrying out the act or series of acts in question. The fact that I do on occasion carry out an act of this type, such as going through a stop sign when the way is obviously clear and no one seems to be around, lends support to the use of "freedom" in this sense. We want to say that anyone was free in some sense to do whatever he has done, and if he has done something which he was not formally free to do, then his freedom would seem to consist in the possibility of his doing it, as evident in the fact that he had done it. Thus, for example, Sartre notes that Orestes' murder of Clytemnestra and Aegisthus, an illegal, immoral and unconventional act, was an expression of his freedom. (Of course it is well known that Sartre means more than "possibility" by "freedom", but this is an essential part of his "freedom".) An act is possible for me if I have the ability to carry it out and if any necessary means for its performance exist. It would seem appropriate, then, to refer to freedom in this sense as my *material* freedom. It would consist of all my possible acts, whether or not they were permitted acts. It would include such acts as smoking "grass" and innumerable others.

If we assume, however, that I am generally a law-abiding, moral, convention-regarding citizen, conscientious in responsibilities and respectful of rules and procedures in my institutional relations, the material freedom to carry out an act without the formal freedom to do it would have only potential value or significance for me, just as the formal freedom to carry out an act would have only potential value for me if it were impossible for me to perform it. Consequently, at this moment, for example, neither my formal freedom to play a sitar nor my material freedom to smoke "grass" is of any considerable significance to me, since the one is impossible for me and the other is not permitted. Of course, I might become concerned with removing each of these barriers if my desire to perform the two acts should gain significantly in strength.

If I should decide that I *must* play a sitar and smoke "grass", and if I should remove the two barriers by learning to play a sitar and by bringing about the legality of smoking "grass", then each of these acts would become a part of my freedom in still a third sense. Or, each act would become a member of a third class or sub-class of acts, namely those which are for me both possible and permitted. And, again, this third class of acts is the "content" or "meaning" for another common use of "freedom". To say that I am "free" to perform an act in this sense is to say that I am permitted to perform an act which is also possible for me. Here one may be referring primarily to the permission, with the possibility taken for granted. But this use, for the sake of clarity, must be distinguished from the use which points only to the fact that an act is permitted, that is, to formal freedom, and says nothing about the possibility of its performance. We may make this distinction by referring to freedom in this third sense as the person's *actual* freedom. My freedom to eat grass, since it is both possible and permitted for me, belongs to this class of my acts and, again, as in the case of the former kinds of acts, there are innumerable others which go to make up my actual freedom.

I have chosen the eating of grass, however, to illustrate this use of "freedom", in order to make a further distinction within the area of my actual freedom and, consequently, to point to still another use of the term, "freedom". One of the "essential" facts noted earlier was the fact that I have no desire to eat grass. I should not, therefore, become unduly disturbed if the government should pass a law forbidding it. I should not care one way or the other. However, there are acts in the area of my actual freedom about which I do care. Playing a guitar is such an act. Although I would not defend to the death my freedom to play a guitar, I should be somewhat disturbed if it should be forbidden by law, or otherwise, and I should be willing to engage in illegal civil-rights demonstrations to have such prohibition repealed, and guitar playing restored to its place as part of my actual freedom.

But guitar playing is more than merely a part of my actual freedom, which it has in common with my freedom to eat grass. Guitar playing is a part of my desired, or generally-desired, actual freedom. And again common usage distinguishes this area from the area containing the innumerable possible-permitted acts of my actual freedom which I have no desire to perform. It is quite common, that is, to think of freedom in terms of the absence of cultural hindrances or barriers to those of our possible acts which we desire to perform.

This is essentially the conception Hobbes arrives at in his definition of a "Freeman". After stating that "Liberty, or freedom, signifieth, properly, the absence of opposition," Hobbes says that "according to this proper, and generally received meaning of the word, a Freeman, is he, that in those things, which by his strength and wit he is able to do, is not hindered to do what he has a will to." (*Leviathan*, pp. 136-37, Oakeshott edition, Blackwell.) Hobbes is speaking here, I think, of the common usage noted above. And to distinguish this from the other three uses noted, the formal, material and actual freedom of the person, it would seem appropriate to refer to freedom in the Hobbesian sense as the person's *effectual* freedom. Hobbes's "Freeman" then is one who has the actual freedom to do what he wants to do.

But Hobbes does not note, or does not recognize, or does not accept, as the case may be, the other uses of "freedom" which I am noting in the terms, "formal", "material" and "actual". Hobbes, it seems, would regard material freedom as "power" and formal freedom as the measure of the absence of social restraints on one's desire-possible acts. But putting all this together the result is effectual freedom as I have described it here, and Hobbes's not noting these other common uses could be either an indication that they were not common in his day, or that he was more concerned with setting ordinary language straight than he was with clarifying it. Or, possibly something else.

But not to be aware of the uses and of the distinctions is to be confused, and not to make them generally in discussion leads to more confusion. This is particularly true in the area of efforts to measure or to compare the extent of the freedom of one person relative to another in a particular socio-cultural situation, or of persons in different situations. The confusion may lie at the outset as to the kinds of freedom being measured or compared, whether one is measuring the person's formal freedom, his actual freedom, or his effectual freedom, and whether, in comparing, one is speaking in each case of the same kind of freedom. And even if this is clear, there may be differences with regard to factors considered in the measurement, as well as, in the case of effectual freedom, how, or with regard to what, it is to be measured. For example, one may measure formal freedom either solely in terms of the governmental or legal limits and controls on action, or he may include the economic system and its rules and regulations, along with other institutional and organizational channels and limitations of action, as factors which must be considered in any realistic measurement or comparison of formal freedom, since these also have significant powers of penalty, and control the access to many of the means of action. And one might add still further to these formal freedom factors the influence of convention and morality with their significant ways of imposing penalties for deviance from the norms. If these are considered along with the legal or governmental controls, then personal formal freedom does not vary indirectly with the extent of legal and governmental controls, as is frequently assumed and argued. An increase in governmental control may, instead, create an increase in formal freedom, by removing other more limiting restrictions on the person's action. Then, further, if one is measuring or comparing actual freedom instead of formal freedom, a whole additional set of factors, including the abilities of the person and the variety of means provided by the social and natural environment, must be considered. And the complexity is compounded still further in any effort to measure or compare effectual freedom.

One note of confusion in the efforts to measure or compare effectual freedom from one person to another arises out of the fact that it may be measured in either of two ways, and that frequently these ways are not clearly specified or distinguished in the comparisons made. My effectual freedom, for example, may be measured either in terms of the ratio of coincidence of my desired acts with my actual freedom, or in terms of the numerical or actual extent of my desired-possible-permitted acts. It seems that Hobbes had the former method in mind. A "Freeman" for Hobbes was one who was not hindered from doing all of his possible acts which he wanted to do. Such a person would have complete effectual freedom. But it should be noted that to measure effectual freedom in this way is not to indicate at all the extent of the person's formal, material or actual freedom. Measuring effectual freedom in this way, one could have complete effectual freedom in a prison cell or a monastery. In such circumstances one would need only to adjust his desired acts to his actual freedom. He would "condition" himself, perhaps through concentration and techniques of self-discipline, towards the elimination of desired acts not possible or not permitted for him. This is, in part, the ascetic way to freedom, and it is, in part, the kind of "complete" freedom he achieves.

But to measure effectual freedom in this way tends to do violence to our more ordinary conception of its measurement. Ordinarily we should think something wrong with our way of measuring any kind of freedom involving action if a prisoner might have more of it than a playboy-at-large.

The more objective and, I think, more common conception of effectual freedom measurement is to conceive it in terms of the overall quantitative or numerical coincidence of the person's desired acts with his actual freedom. Thus, the person with more desired acts coincident with his actual freedom would have more effectual freedom than the person with less extensive coincidence even if the latter had a higher ratio of coincidence and consequently, perhaps, a lower level of frustration. (We tend ordinarily to separate "freedom" and "frustration" in granting that a person can be relatively free yet considerably frustrated. Frustration usually is associated with the intensity of a particular unfulfilled desire.) Furthermore, measured in this objective way, my effectual freedom is more accurately comparable to the effectual freedom of others and I can speak with more precision about methods of its increase and ways to maintain it.

I have suggested that my desired acts are the most significant factors in the quantitative extension or measurement of my effectual freedom when I think of its measurement in the more subjective terms of the ratio of coincidence of my desired acts with my actual freedom. And, as noted, I can have complete effectual freedom, measured in this way, or become, in other words, Hobbes's "Freeman" without altering anything in my external circumstances. It would seem, consequently, that when my effectual freedom is measured in the more objective way of numerical extent, my desired acts would become less consequential for the increase of my effectual freedom. This is the case, but they remain, nevertheless, a significant factor. I may, for example, increase my effectual freedom, objectively measured, by increasing my desired acts along lines coincident with my actual freedom. To do this I may either develop interest in known acts which are possible and permitted for me, or, I may, through investigation, education, etc., discover or learn of acts which are possible and permitted and which, immediately upon discovery, I find desirable. With regard to the former method, I know, for example, that it is possible and permitted for me to play golf, that playing golf is now a part of my actual freedom. Consequently, should I look into the matter further, perhaps try a game or two, I might develop a desire to play golf, and it would become then a part of my effectual freedom along with guitar playing, and my effectual freedom would be increased to that extent. As regards the second method, everyone has had the pleasure of discovering some act, or kind of act, permitted and possible for him in his circumstances, which he found, immediately upon its discovery, to be desirable and, consequently, through the discovery alone, brought about an increase in his effectual freedom.

Thus, in these two ways, and still in terms of my desired acts only, I may increase my effectual freedom without purposefully bringing about any change in my actual freedom.

On the other hand, I may choose to pursue an increase in my effectual freedom by a direct attack on my circumstances aimed at altering or increasing my actual freedom to bring it into coincidence with one or more of my desired acts. Since my actual freedom consists of the coincidence of my possible acts with my permitted acts, that is, of my material freedom with my formal freedom, I may seek to increase it either by making possible an act already permitted for me, or by gaining permission for an act already possible for me. Going back to a former example, I could learn to play a sitar, or I could work to bring about laws which would permit the smoking of marihuana. And there are, obviously, other ways in which I may alter either of these areas of my freedom, such as creating means to coincide with my present abilities and thus bringing about additional material freedom; or, I may remove myself from one set of circumstances to another, leaving, for example, one organization where something is forbidden and going to another where it is permitted; or, I may remove myself entirely from one society to another.

Still another method of gaining an increase in my actual freedom through an increase in formal freedom is found in the economic system or structure of most societies. Money and the property system, supported by a complex network of rules, procedures, organizations and laws, is one of the major methods of controlling access to the existing means for action in the society. It is, consequently, along with custom, morality, other organizations and government directly, a part of the permission pattern or formal freedom structure of the society. Consequently, a mere increase in my income constitutes an increase in my formal, and therefore actual, freedom. It serves as additional "tickets of admission" to the means for action provided by my natural and cultural circumstances. The existence of concerts, for example, and my ability to go constitute my material freedom to attend concerts, but without the money to purchase a ticket (and assuming that I have no credentials or "connections" for free passes) I do not have the actual freedom to attend a concert, nor, of course, the effectual freedom to do so.

If I think in terms of the long-run maximization of my effectual freedom, the maximum actual freedom would provide the most favorable circumstance for its realization. That is, the greater the extent of the kinds of my performable (both permitted and possible) acts, the more favorable the odds on their coincidence with my desired acts. Consequently, in seeking the long-run maximization of my effectual freedom I should seek the maximization of my actual freedom, that is, the maximum coincidence of my possible acts with my permitted acts. As noted already, I may approach this by alterations or additions in either the one or the other, or both. But there is an intimate relationship between my material and my formal freedom which makes it impossible for me to consider them in any long-run fashion as independent entities. One reason for this is the fact that other persons function prominently as determinants of both kinds of freedom for me. Other persons are factors in my formal freedom as the holders of the power of penalizing me for deviance and as the source of origin of the norms. Other persons are factors in my material freedom as a two-way source of the means which make possible many of the acts I am able to perform. They provide means for me as the productive agents of many of the tools, instruments, etc. which I use in action. Someone's making a guitar, for example, is essential to my material freedom to play it. Other persons also supply means more directly in contributing to my material freedom. They do so when their abilities become the essential means for certain acts in which I may engage. Concert-going, for example,as a part of my material freedom, depends directly on the concert artist's ability to perform. His ability is a direct part of my means. In the same way, the teacher's ability to teach constitutes an essential part of my material freedom to learn and to develop many of my own abilities. And so on —obvious examples of this kind are numerous.

Thus the extent of my own material and actual freedom depends both directly and indirectly on the material and actual freedom of other persons, and the maximization of my own would seem then more likely to be achieved under the maximization of the actual freedom of all.

The greatest actual freedom of the greatest number would require the maximization of the material freedom of the greatest number, and the extent to which this could be achieved would depend on the nature of the formal freedom structure, that is, on what is permitted to what extent for whom.

Consequently, in the promotion of my own maximum effectual freedom I should seek to establish and support that formal freedom structure or system which tends to create the maximum material freedom for the greatest number of people. Or, where there are the most abilities of the most people creating and constituting the most means for the development and expression of my own abilities I will be most likely to have the maximum of effectual freedom. Which formal freedom structure is most likely to achieve this in any particular situation will depend, of course, on the situation, that is, on the present state of abilities and means.

Randolph-Macon Woman's College,
Lynchburg, Virginia.

The Right Answer to Pontius Pilate

by H. S. Harris

Foreword: A note on the form of what follows

What follows is not only an argument but an illustration or exemplification of what is argued for. It is also, secondarily, a philosophical interpretation of the technique of argument in medieval philosophy.

The basic position underlying the argument here stated is that rationality, the distinguishing characteristic of *homo sapiens* is a social product and that man could not be a rational animal if he were not already a social one.

Thus *thought* the distinguishing activity of *homo sapiens* is minimally dialogue and Plato gave the first correct definition of it when he defined it as the dialogue of the soul with itself. He also was the first to put forward something like the definition of truth here offered (*Gorgias* 472b-c; cf. *Gorgias* 486e-487a and the whole subsequent argument). But since he also recognized as a necessary consequence the impossibility of *completing* a dialogue in any positive sense (no real agreement between Socrates and his "touchstone" Callicles is possible) he saw that the purpose of *writing* philosophy can only be to *stimulate* discussion.

But to write a dialogue is not the best way to start one. One must somehow *show* the object of rational discussion —one must reach a conclusion that does justice to all viewpoints— without claiming finality for it. The medieval technique of Question, Objection, Response, and Solution of Objections, seems ideal for this purpose. This technique is of course only a formalization of the dialectical method not of Plato but of Aristotle (exemplified for example in his discussion of the question "What is the good for man? " in *Ethics* I, 5-12, where chapters 5 and 6 are "Objections", Chapter 7 is "Response" and chapters 8-12 are "Solutions"). Plato's method emphasizes the theoretical endlessness of inquiry and Aristotle's the practical ending of it. What was needed here was a compromise between the method of Plato and that of Aristotle; and this has been achieved by a novel use of the ancient device of the explanatory *scholium* to provide *ad hoc* answers to the objections. The device was not formally employed as far as I know by medieval thinkers in anything like the way I have used it here but they did at times discuss the objections as they were raised while reserving formal replies until after the Response. Thus I believe I have remained close both to the medieval thinkers and to Aristotle in all the essential aspects of my mode of argument.

"The Right Answer to Pontius Pilate"
Fragment of An Apocryphal *Quaestio Disputata de veritate*

(The question appears to be "What is Truth" but the introductory division and everything except the first Article is lacking).

First Article

Whether there is anything that we *do know* for certain to be true; or (alternatively and more accurately) whether there is anything that we *must claim* to know beyond doubt and beyond the possibility of effective contradiction.

It would seem that there is not: FOR

Obj. 1. No empirical matter of fact is true beyond doubt and beyond the possibility of effective contradiction.

Scholium This objection I take to "obvious" in the present context.

ad Obj. 1. But it requires to be supplemented by:

Obj. 2. Purely formal statements of logic and mathematics are not true beyond doubt and beyond the possibility of effective contradiction. This objection is probably not "obvious" in the present context so I will defend it thus:

FOR: A logically true statement is "true" only within the system if it is true at all. There are some reasons for saying that we ought not to say that logically valid statements are "true" at all: for instance the notion of logical validity is not confined to simple two valued systems where all statements are either true or else not true but false.

Scholium We can argue that this is on the contrary the very reason *(ad Obj. 2.)* for saying that the two valued logic of consistency with respect to negation is "true". For this logic we might say *is* the logic of our concept of truth itself. It formalizes just what we mean by *true* as opposed to *false* and since there are so many possible logics it is all the more important to know and to be able (and be permitted) to say what is *true* (as distinct from what is false or *not true*) about our concept of truth itself. *Have we then discovered here a truth beyond doubt?* (Perhaps we have but when the consequences of it are formulated clearly they do not seem to be what Pilate was asking for). In any case *it would seem that we have not* for a further objection here arises:

Obj. 3. There are in fact several different theories about our notion of truth itself. Thus there is no absolutely certain truth about our concept of truth. It seems likely that none of the theories so far put forward (say the "correspondence theory", the "coherence theory" and the "pragmatic theory") is "the simple truth" about truth and it is surely possible (-conceivable within the two-valued logic of our concept of truth?) that there is no simple or final truth about truth at all unless it be that our concept of truth is inconsistent and involved in contradiction (this would be an absolute truth of a kind but one that Pilate did right not to wait for beause he would doubtless have said that he was quite well aware of *that* already).

Scholium This objection in its turn can be met by the argument that one of *(ad Obj. 3)* the theories of truth *—the correspondence theory— is at least part of the truth about truth* because all other theories of truth that are even faintly plausible contain it in some form as a necessary presupposition. This answer to the objection can be allowed to stand for the moment because it leads us directly to the crucial objection which reveals the weakness that makes an answer to Pilate in terms of formal logic unacceptable. This fourth objection will be the final one to be considered here:

Obj. 4. Even if the correspondence theory be allowed to stand as absolutely true it does not resolve Pilate's question, for there are, as a matter of observable fact, many theories which "correspond" with "reality" in different ways which are all sufficiently specifiable to allow the employment of the concept of "truth" in all of them, even though the theories themselves may be logically incompatible with one another in various ways: and it is therefore reasonable to suppose that "reality" is in its nature patient of variant interpretations none of which can claim privileged or absolute status. Indeed this hypothesis seems to be the only one upon which an unbiassed appraisal of Pilate's real problem (about the "true" destiny of man) can be based. Therefore it would seem that the question he posed is logically unanswerable, that he posed it precisely in order to compel recognition of this fact, and that his going out without waiting for an answer was the best way of enforcing the point that his question is not one to which a *true* answer can be given.

Scholium This objection is the work of a real demon (not of any philosopher's fancy). *(ad Obj. 4.)* Like all inventions of the devil it takes many forms and for this reason it cannot be finally refuted (although the conclusion is demonstrably false no matter how it is arrived at because if it can be *arrived at* then rational communication is possible, and if rational communication is possible it is the appropriate human aim presupposed even in the raising of the problem about human destiny, and hence the conclusion is false). Furthermore, like all the inventions of the devil the objection contains nothing but a cunningly distorted image of the truth. This has been made as clear as possible here by stating the argument and the conclusion in their most dangerous form —the form in which they are made to appear necessary presuppositions of any achieved brotherhood of man. For certainly it is true that we must abandon all simple linear philosophies of progress if we wish to obey the Second Commandment —"Love thy neighbour"— or the second formulation of the Categorical Imperative —"Act so that you treat humanity, whether in your own person or in that of another, always as an end and never as a means only". Does it not therefore follow as a corollary that we must adopt Pilate's attitude to the question he asked? The answer that he did not wait for is dramatically obvious in the context of the Gospel. Jesus has just asserted that "for this cause came I into the world that I should bear witness unto the truth" implying that the truth is something independent of his witness. But he adds at once "Everyone that is of the truth heareth my voice" and of course his confidence of this rests on that other claim recorded by St. John "I am the way, the truth and the life". St. Thomas would probably have prefaced *his* reply to the question by citing this authority perhaps in combination with the saying of the Philosopher "To say of what is that it is not, or of what is not that it is, is false, while to say of what is that it is, or of what is not that it is not, is true". But the cumulative result of the above objections is to make this appeal to Holy Scripture appear sinful and the appeal to reason foolish.

SED CONTRA:

On the contrary, the Apostle says "Ye *shall* know the truth and the truth shall make you free"; and the Commentator (- C.S. Peirce as the *Philosopher* herein is Kant) rightly declares "The opinion which is fated to be ultimately agreed to by all who investigate is what we mean by the *truth*, and the object represented in this opinion is the *real*". Likewise it is written "Seek and ye shall find: knock and it shall be opened unto you".

RESPONDEO, DICENDUM EST QUOD:

I answer, it *must* be asserted that we know beyond doubt and beyond the possibility of effective contradiction that when men inquire together sincerely about any matter in dispute between them they *can* in the end find a solution that is seen to be satisfactory by all of them and by all rational observers. In the process of inquiry the problem may be radically transformed, but the term "satisfactory solution" does not include the "giving up" of the problem. To say this is only to say that men can achieve rationality. To attempt to deny *this*, to utter the formula "It is not absolutely certain that men can ever achieve rationality" is only to utter a string of sounds which are demonstrably falsified by the act of understanding them or claiming to know what they mean. Any one of us can give a meaning to the assertion "You are not being rational" when it is directed at him by someone else, and thus we may come to believe that we can conceive what would be meant by the assertion "No men can ever be rational". But there is only one imaginable circumstance in which we could say this and know what we were saying, and that would be if something like the Christian vision of the Last Judgement came true. But then *ex hypothesi* we should be miraculously lifted out of our present situation of seeing through a glass darkly. We should all be at that moment rational and the *fated* agreement of all inquirers would come to pass. The sentence would be false and the most that could be true would be "No men have ever been rational and I now see that no man ever can be rational except by the grace of God". But, that is *not* our present situation and even if we supposed, as I for one do not, that that is how the ultimate agreement is fated to come about, it is still the case that the only rational assumption in our situation is that however dark the glass may be we can see through it if we try long enough and hard enough. If this is not the case we cannot even intelligently *say* or suppose that it is not the case because we cannot intelligently say or suppose anything. It may nevertheless be true even though we cannot suppose it (and hence the more accurate statement of the problem announced at the outset is concerned with whether there is anything that we *must claim to know beyond the possibility of effective contradiction.* No *man* can *effectively contradict* the claim that rational communication is possible for men even if perchance it is not. We might for example really be playthings of Descartes' malignant demon —or, more realistically, of evolutionary processes— but it would make no more sense to say we were *deceived* on the first hypothesis than on the second. We would not on this hypothesis be capable of being either deceived or undeceived.

I suppose —indeed I very much hope— that I am merely flogging a very dead horse here. The part of my thesis that will, I surmise, seem most open to effective contradiction is the assertion that when we say we must always be able to arrive at a universally satisfactory conclusion we are saying *no more than* "It is possible for men to be rational and to communicate their differences of opinion to one another and inquire into them rationally". Yet this equivalence must hold if we are to be able to distinguish rational communication from non-rational communication (which certainly occurs) and more rational from less rational modes of inquiry (I do not think anything strictly describable as non rational inquiry can occur —unless tossing a coin is *inquiry*— it is a quite respectable outcome of inquiry but that is different).

Rational communication is any form of communication which is both aiming at rational agreement, and employing a method consistent with the achievement of that aim. Rational agreement about any problem is essentially agreement about what constitutes a disinterested standpoint or a neutral analysis *of that problem.* Nothing at all can count as an *analysis of the problem in which* the possible outcomes with the paths by which they can be achieved are not identified, and nothing can count as a neutral analysis in which these outcomes are not evaluated from the point of view of all parties interested in the dispute (including those who may be absent from the discussion).

It may well seem that this ideal of rational agreement is never achievable; and in one sense this is true. It is true by definition that (setting aside the Day of Judgement hypothesis) the finally conclusive agreement can never be reached; no matter how well we may come to agree or how secure we may feel about our methods of achieving neutrality in analysis of disagreements we can never know that we are *fated* to remain firm in our agreement (because fate here is simply a name for what actually happens and what actually happens is only known when it happens). The truth that makes us free is the truth still to be discovered in the future, and the only truth we can be *absolutely* sure of is that this *must* be so and hence that we must always be concerned to preserve this situation in order that we may have reliable standards for distinguishing between truth and error at all.

It can readily be seen that in this answer the crucial conception is not that of truth but that of *rationality.* Truth has been defined as a purely regulative ideal for communication and inquiry. Thus one might say that it has been demoted in the realm of values. Pilate has been answered but at the same time any literal interpretation of a claim such as "Jesus is the Christ" has been denied. It never *could* be true that "The Word (of truth) *was* made flesh" or that any sentence in the past tense asserting the complete realization of the truth could be true. Of course anyone is at liberty to believe that *rational humanity* has been instantiated in some man now dead, and he may believe that "Jesus is the Christ" in that sense; but if he believes this then he is either committed to believing that any man can be the Christ and hence that it is only natural that other communities should find rational humanity instantiated in other models, or else to admitting that Jesus was not really a *man* at all and hence that his appearance was irrelevant to our common human problems.

Properly speaking, however, it appears to me to be a fundamental error to suppose that rationality can be instantiated in a single mortal individual of the species *homo sapiens* at all. Perhaps this is the most fundamental root of error, in so far as error can be supposed to have any rational structure at all —for there is nothing necessary about the genesis of errors as there is about the genesis of truths, and the important determining causes of errors are not rational at all but derive from the needs, impulses, wishes and practical concerns of particular human animals or groups of human animals. I call it the root of error because it is the belief that makes possible the establishment of authorities that are independent of rational discussion and rational inquiry; and any authority that is thus removed is (we might say by an appropriate transformation of the definition of truth) "fated in the end to become irrational".

In order to meet objection 4 in all of its multifarious forms precisely what is needed is the deliberate and conscious abandonment of the idea that the rational man is or can be a thinking *substance* —an entity possessing in itself the foundation on which truth can be established— and the deliberate and conscious acceptance of something more like Pascal's conception of man as a thinking *reed*, a thing sensitive to every wind of opinion but possessing a shape and life of its own and not patient of any and every shape or any and every pressure, something that grows in clumps and is capable of surviving and repairing almost any damage done by the violence of some force other than the wind as long as it is rooted in its native habitat among the other reeds, but quite unable to withstand even a sudden change of the wind when isolated from the rest. It is in and through communication with others like ourselves that rationality, the capacity to think, is brought to birth; and it is in and through the establishment of rational communication (as distinct from emotional involvement which is the more primitive, typically *animal*, type of communication) that *homo erectus* —already a being who dwells out of natural necessity in communities becomes *homo sapiens*— the being who can consciously establish communities and can deliberately set out to establish the rational community of the species as a whole.

When rationality is thought of in this way, rational thought becomes itself a mode of public social interaction and behaviour which can be criticized and evaluated publicly by standards and criteria that are universally acceptable without reference to the private experiences, intuitions, certainties, doubts, confused notions or clear and distinct ideas of any single individuals whatever. But it should not be thought that either the existence or the importance of private personal experience and conscious life is in any way undercut or undervalued by this. On the contrary it can be quickly shown that in the view here advanced it is the spontaneity and free play of this personal private experience that is assumed as the criterion of rationality itself and hence as the absolute value. If what is absolutely true must be determined by reference to the logical presuppositions of rational communication, what constitutes rational communication is determined by the criterion of maximizing personal freedom and guaranteeing the inviolability of privacy. Thought itself as the establishment of a standpoint valid for all parties to a problem must be conceived as a mode of *social action* which is of overriding importance in *public* life precisely because the fundamental problem of a rational community is to minimize the appeal to irrational, involuntary interactions between its members. It is a fundamental assumption therefore that everyone should believe what he chooses (or rather that he should be moved to belief only by the free blowing winds of opinion). But the expression of his belief in public, the modes of his thought (i.e. of public argument) and of his action must be regulated by the canons of rationality. On the other hand no man can have the freedom to remove himself altogether from the climate of rational inquiry which the necessities of community life generate. Thus there is nothing equalitarian about the freedom of opinion therein guaranteed. Any man may believe what he chooses. But if his opinions themselves conflict with the presuppositions of rational communication (as those of professed Nazis do for example) he will be at an inevitable disadvantage. He cannot be tolerantly allowed to express his beliefs in the only way in which they can *consistently* be expressed (i.e. in *deeds*); he must express them if he wishes to publish them at all in the mode of public argument, a mode which is radically inconsistent with the beliefs themselves.

To take a less obvious example: when convinced Marxists or Freudians (whose beliefs are perfectly compatible with the maintenance of rational communication, and indeed are aimed at establishing it on sounder foundations than hitherto) begin to answer the arguments of their adversaries by explaining how in *their* view these people come to have such erroneous opinions we must immediately recognize that they are in the wrong. They are necessarily in error even though it might turn out that their view was right and their analysis of the error was right, because before we can properly consider the *genesis* of an error we must establish the neutral standpoint from which all parties can perceive that it is an error that is under discussion. No mode of expressing one's beliefs which ignores the fundamental requirement of neutrality in inquiry can establish the truth of the belief even though the belief may in fact turn out to be true when someone else sets about the task of establishing it properly.

REPLIES TO OBJECTIONS:

To Obj. 1. Given a clear agreement about the definition of an actual problem and the appropriate method of its solution the solution arrived at by appeal to the facts often is true beyond doubt and beyond the possibility of effective contradiction. Two friends who wish to go fishing can properly rely on the tide tables and the sunrise or sunset times provided for the use of anglers, while two astronomers interested in the motions of the sun and moon may dispute about the accuracy of the tables, and two philosophers may even dispute about the *kind* of certainty that the statement "the sun will rise tomorrow" possesses.

To Obj. 2. It is always necessary in any rational dispute to settle what the criteria of a satisfactory solution are; and one part of this is agreement on the logical canons to be employed. But there is nothing really absolute about consistency with respect to negation because there is nothing absolute about *any* verbal or symbolic formulation of a solution. This is partly because, as Aristotle already saw, different types of inquiry admit of different degrees of exactitude, and partly because the conceptual structures in which we express our differing points of view are very flexible so that views that may appear to be logically incompatible are often reconcilable in fact when they are correctly understood. The real logic of inquiry is the logic of correct understanding; consistency with respect to negation is generally a useful instrument in achieving understanding, but it has no peculiar status or absolute value. (This is not to be taken as an apologia for any "non-Aristotelian" logic; the instrumental logic of all human dialogue is based on Aristotle's "laws of thought" and it is quite wrong to speak of "dialectical logic" as if it was an *alternative system* that anybody could ever *use*. Dialectic is only a different way of making use of the ordinary two-valued logic. Other logics might become useful in certain circumstances but as far as I know they are not presently employable in human communication. (See further the Reply to Obj. 3 below).

To Obj. 3. The pragmatic theory of truth is in all essentials the *true* one although for reasons indicated above and below no absolutely true (i.e. final) formulation of it in the language of discourse is possible. It is true precisely because it provides a logically satisfactory explanation of this impossibility of final statement. The defence of the absolute truth of the correspondence theory given in the Scholium is inadequate because the way in which it is contained in other theories varies. For example, if the coherence theory is taken to be true, then it may be absolutely or perfectly true that truth is *adaequatio intellectus et rei.* But it is true with equal absoluteness that we can never know anything perfectly (i.e. in the mode of exact correspondence) except the definition and the impossibility of producing a non self-referent instance which satisfies it. This is the dialectical shipwreck of the correspondence theory (and of the coherence theory too); the only way out is to give up both the ideal of perfect correspondence and the ideal of perfect coherence —or to put them off into the indefinite future, as in the pragmatic theory of which the response above is a form. Thus in fact Obj. 3 is valid against taking two valued negation-consistent logic as *absolute* truth —it should be taken simply as the present instrument of rational discourse. It may in fact be indispensable, or it may not. If there were beings who discoursed about their experience in a multi-valued logic and *if* we somehow came in contact with them, I see no reason for ruling out *a priori* the possibility of mutual communication. But of course these are enormous *ifs* and there may be something empirically or logically wrong with one or both of them.

To Obj. 4. The premisses of this objection are true but the conclusion is not valid. The replies to objections 1 and 2 indicate that any use of the concepts of empirical and logical "truth" is dependent upon certain agreements about the aim and purpose of the common activity. Because life can be lived in different ways it is to be expected that different standards of truth and relevance will be employed by different groups that live or work together. Nevertheless it does not follow that there is no absolute truth because it does not follow that there is no purpose that all human communities must share. There is in fact such a purpose for any community which has a concept of "truth" as something to be inquired about and discovered (and no concept that is not analogous to our "truth" in these ways can be regarded by us as a concept of "truth" at all). This purpose is the maintenance of rational communication and the minimization of all kinds of arbitrariness and violence in the communications of its members with all other beings capable of communicating rationally.

It must be realized that the absolute commitment involved in accepting this purpose as fundamental does not (or need not) involve belief in any final system of truth. We are committed to believing that "for *any* problem that can be stated a rational solution can be found" not to the quite different belief that "a solution can be found for *all* problems" (as if they could all be stated and solved at once which is logically impossible on the view here taken). Even rationality in the view here stated is essentially treated as an instrumental value. To be precise it is regarded as the methodological aspect of *freedom* (by which is meant the conscious and deliberate experimental discovery and realization of human potentialities in public life and public objects).

There *is* a sense in which the requirement that all human activities must be methodologically rational commits us to belief in the unity of human nature and hence in the unity of truth. But the way in which *una veritas in variis signis varie resplendet* (Nicholas of Cusa) is only subject to the curb of logical or systematic consistency very indirectly. Assumptions that are in some sense logically incompatible do not *need* to be reconciled unless and until they give rise to practical problems in human communication and action. It is easy to see why scientific inquirers into human nature are bound to go on making assumptions in different fields without being able either to reconcile them with the assumptions made by those in other fields or to convince all concerned that the assumptions made by others are mistaken. What is problematic determines what can be assumed: and there are different ways in which human nature in society may be (or plausibly appear to be) subject to control and change. Practical difficulties only arise if the use of one method of discovery and control (say in economics) produces evils which are revealed by another (say in psychology). At this point a reconsideration of assumptions becomes *necessary* and the presence of theoretical incompatibilities may well be a useful indication of where to begin. Of course it is always irrational (at least in the sciences that deal with man) to declare with finality that some *specific* theoretical incompatibilities have to be accepted as ultimate.

Or again, it is typical of scientists to be tormented by problems that appear to have no practical relevance and it is radically irrational to deny anyone's right to seek a solution for any problem or to assert *a priori* that there are aspects of experience that are beyond the range of scientific investigation. It is not certain that there is anything that we cannot know. But nonetheless it is certain that there are some things that we *must not do.* For instance, there are experimental procedures which *cannot count as scientific,* if "scientific" is taken to mean "appropriate for use by a rational inquirer —e.g. some of the "experimental" work conducted on human subjects by Nazi medical researchers. "Experiments" of this sort do provide data for the rational inquirer, but only in the way in which observation of the course of nature can provide data even where it is not *physically* possible to reproduce the observed phenomena under laboratory conditions.

The same considerations apply to cultural differences. We need not hesitate to condemn cannibalism or head-hunting and to insist that these practices must somehow be brought to an end, although of course the problem of how this can most rationally be done will need careful study, and will give rise to disputes that can only be resolved by experiment (and hence) cannot be resolved finally at the theoretical level, since a recreation of the conditions for purposes of further experiment would be illegitimate). To take examples from a culture which could not by any neutral standards be called more backward than our own the British were right to condemn and suppress the religious practices of thuggee and suttee in India (although quite possibly they were not very rational about how they did it). But when we come to such cultural differences as polyandry, polygamy and monogamy (to take a set of hardly compatible alternatives that have critical import for the value structure of any human community) it is not at all obvious that any one system of social organization is the human or rational one, though it *is* clear that every form of family organization has its own weaknesses and its own characteristic diseases. It appears, at least to this observer, that the only position of rational neutrality here is one that accepts cultural pluralism and simply seeks to obviate the difficulties (of moving from one culture to another for example) that arise from it, and to rationalize human relations within each traditional "way of life" as far as possible. *Perhaps* by doing this we *may* eventually create a cultural melting pot even for this pluralism. But I am inclined to doubt it; and because this is, as far as I can see, the only *rational* path towards an answer to the question "What is the best form of family organization? " I am inclined to say the problem has no solution because there is no way to give a univocal yet neutral meaning to the word "best" in the question. It does not make sense, therefore, in this area to ask "What is the *best*? " though it does make sense to say (rightly or wrongly) "This way is *certainly bad* and that way is *probably* better." The *one truth* of human nature shines *back* just as clearly in the relative and comparative form of the positive judgement as it does in the absolute form of the negative one.

Glendon College, York University,
Toronto, Ontario.

The Justification of Morality

by L. L. Haworth

Following the argument developed here will be a bit like taking a train trip when you aren't certain of your destination. From time to time the argument like the train will grind to a halt and there will be an opportunity to debark. One may feel that he can go so far with me and no further. I shall try to provide reasons for going all the way, but the argument may seem less plausible the longer it goes on, and so at the stopping points along the route an effort will be made to convince those getting off that even so the trip was worth their taking.

The first station is that at which one has become convinced that "we ought to be objective." At the second, one will have decided that the principle that we ought to be objective is not merely formal but substantive, in the sense that from it and discernible matters of fact we may on occasion infer differential moral judgments —judge, that is, that some lines of action ought to be preferred to others. And at the third and final station one will understand that so substantial is the principle that on it one may and ought to found the whole notion of morality, a conclusion that has as I believe extensive implications for the conduct of life and the structure of society, for the meaning of death —the meaning with which we ought to invest it— and the nature and significance of freedom. In all these matters the resulting position seems mildly Kantian.

I

We ought to be objective. Why? The place to start is with the meaning of objectivity and its synonyms, "disinterestedness" and "impartiality." One is objective in solving a problem insofar as he is open to any matter relevant to its solution, is willing to entertain any relevant factor and to settle the problem in a way that reflects the kind of relevance the factor has. Being objective means undertaking to settle issues "on their merits." The image is of the decision maker, so far as he is objective, as a special kind of machine —think of a canning factory— that takes in raw materials (relevant evidence) and, by working these over in some fashion that needs to be understood, puts out decisions that in no way reflect the characteristics of the machine (if the output is prepared food then machine oil isn't present in its taste) but are formed wholly by the evidence it has been fed. Its decisions are nothing more than a sort of distillation of the input. Being objective involves working over the material in this way, but also actively seeking out everything relevant to the problem to which one is addressing oneself. The alternative as a manner of ingesting the data is to bias it, which means to allow an interest or whim of one's own to interfere so that the decision is different than it would have been if that interest or whim had been withheld. There is then a character of impassivity in objectivity. One who is put off by mechanical images may think of the angel holding the scales on which on judgment day are weighed and appointed to their places the sinners and the saved. Their lives and characters offset the scales —not an angelic thumb— and the angel only notices the fact and by pointing up or down translates it into a decision of some moment. Presumably, too, the blindness of justice is its lack of regard for factors in a problem that might otherwise skew a decision —its blindness is its objectivity.

Much more needs to be said about objectivity, and will be. But for the present enough has been said to argue for the principle of present concern. We ought to be objective. That this is so can be seen by reflecting that when we say of an action that it ought to be carried out we are in fact saying that it is the action that would be preferred by one who decided the matter objectively, who let his preferences be regulated by considering the matter wholly and solely on its merits. If a person begins by thinking that some course of action ought to be adopted, and then becomes convinced that in the reasoning by which he defended his view he allowed a bias to enter, which skewed it, two courses are open to him. (1) He may simply give up the claim that it ought to be adopted. In this case he may nevertheless continue to prefer the action, but he will not dignify his preference by saying it "ought" to be chosen. The idea here is that he will shift from imputing a character of universality and objectivity to his preference, to merely announcing that he has it, and perhaps encouraging others to adopt it as well. If he did continue to describe the action as one that ought to be followed, admitting that his attachment to it arises simply out of a personal interest and evincing no belief that there may be a way of establishing the preference without introducing a bias, we would be puzzled or perhaps angry with him. He either doesn't know the use of "ought", or is making sport with us. (2) Or, he may continue to hold that the action ought to be carried out, on the ground that although he does not now know how objectively to support his preference he is nonetheless convinced that it can be in that way supported. He has faith.

If, however, he lacks this faith, is honest, and knows the use of "ought", he will not claim that the action ought to be carried out.

I mention in passing that there are at least two other senses of "ought" that are not relevant here and that might be used appropriately in recommending an action which could not be recommended as the one an objective decision maker would prefer. These are the so-called prudential and the neutral "oughts". The point of the first is to allow one to recommend an act as being in the best interest of the person to whom the ought is applied: If I think that all-things-considered it would be in your best interest to stop reading at this point, I may appropriately say that you ought to do so. Even so, I may not want you to stop; the meaning of the words, unless in the case where I do have a desire that you should stop or continue it seems relevant to take account also of the resulting non-cognitive meanings the words will have, is such that I could equally well express myself by saying that it is in your best interest to stop. A neutral ought, by contrast, is neither prudential nor objective. Nor does it make a statement. When I say that you ought to stop, meaning neither that it is in your best interest nor that your stopping is the course a disinterested decision maker would recommend, instead of asserting something I may be merely uttering the command, "Stop!" (I ignore here further possibilities that it wouldn't be interesting to pursue, as that by "you ought" is meant that this is the course implied by the customs that prevail.)

To mark the distinction, I shall call the "ought" that does concern us here the "objective" ought. Some such thought as this one would be appropriate at this point: Well, there are at least three meanings of "ought", and when it is said that we properly give up describing actions we prefer as actions that ought to be followed when we see that they would not be preferred by an objective decision maker, this applies only to one of the three, namely, the objective ought. But in that case the position is trivial and uninteresting. To say that something ought (objective) to be done is to say that it is the course that would be preferred by an objective decision maker, and being objective means preferring what an objective decision maker would prefer; consequently, in saying that we ought to be objective one is only saying that the course of preferring what an objective decision maker would prefer is the course that would be preferred by an objective decision maker.

Seeing one's way out of this difficulty requires seeing that (a) saying "ought" in the objective sense involves also directing that the action be performed, so that the claim that an action ought to be performed is translatable into (i) "it is the action that would be preferred by an objective decision maker," and (ii) "Do it!"; and that (b) there is a peculiar appropriateness of the objective "ought" to moral puzzlement, so that when people are uncertain as to what they ought to do, the nature of their concern is such that the fact, however trivial, that they ought to be objective is highly relevant. For when they wonder what they ought to do (unless what is unclear to them is where their best interests lie, or which course to follow to realize their own interests, neither of which —as such— forms a *moral* problem), they are raising a question about the objective ought. Consequently, the otherwise trivial statement, that they ought to be objective, is entirely to the point, that is, not uninteresting. In the same way, when one wonders what the product of 7009 and 17 is, that the answer is trivial does not make it a less adequate settlement of the wonder.

Owing to the first circumstance, that the objective ought is a directive, the principle that one ought to be objective is more than the mere tautology that the course of preferring what an objective decision maker would prefer is the course that would be preferred by an objective decision maker. It is not more in the sense of not being a tautology, but in the sense of being something else as well. The directive is significant: there are some courses of action that are consistent with it and others that violate it. Owing to the appropriateness of the analysis of ought contained in the tautology, "One ought to be objective," to moral puzzlement, the directive expressed by the principle that one ought to be objective has point as the directive that precisely solves the puzzle. The puzzlement is formed by uncertainty as to which of a variety of directives to follow, and the sole condition is that the directive chosen shall be the one that ought to be chosen. The directive, "Be objective! ," satisfies the condition, owing to the fact that the principle, "One ought to be objective," is a tautology.

An analogy will help. In a baseball game, if a batter asks his manager whether he ought to bunt or hit away, he is both seeking direction and asking for information as to the action best calculated to enable his side to win the game. Imagine, however, a peculiarly dense player who is not aware that in the context his question, "What ought I to do? ," *is* a request for information as to the action best calculated to enable his side to win the game. Then the trivial answer, that he ought to follow that course best calculated to enable his side to win the game, might have some point. Whether it does will further depend on whether any directives can be inferred from it, in the sense of actions that can be recommended as likely to lead to the desired result. Similarly, when outside any special context, such as a baseball game, one wonders what one ought to do, the (trivial) response that in any case he ought to be objective, is in any case true, and moreover has point just in case there are in the circumstances available and discoverable actions that better fulfill the requirement of being objective than others would. That is, just in case the principle that one ought to be objective is not merely a formal one, but substantive, in the sense that one can derive from it particular actions as being preferable.

There is an additional reason for calling the ought in point "objective". Judgments of what ought in this sense to be done are themselves objective in a distinctive way. This is the Kantian point. If I say that you ought to stop reading, meaning that this is the course an objective decision maker would recommend, my advice purports a character of objectivity that distinguishes it from advice that only expresses my preferences. In saying "I want you to stop, so stop! " the information conveyed is, if true, true as such and in general, hence in a sense objective; but the command, in a parallel sense, is not. That is, so far, the command cannot be regarded as binding on any but myself (or, if one wishes, not binding at all). No reason is given for another following the command. But if it is shown that you ought (objective) to stop, then by that fact the command is detached from the person who utters it, and is caused to stand on its own ground, as having merit in itself or as such, that is, in general, and hence for all alike. If a decision is reached by a process in which the distinctive traits of the decision maker as a person in no way skew his decision —like the machine that doesn't smear oil up the side of the can it packs with food— then the only way another has of avoiding that decision, of showing it to be ill-advised, is by inserting a whim or interest of his own that skews it. And this is "avoiding" the initial decision in the sense of refusing to accept its plausibility, not of showing its implausibility. If the machine packs the cans in the indicated way, then any other machine that addresses itself to the same task must pack them in the same way —or else subjectivize the packing process, dribble oil in the cans, or in general add something of itself to the product.

The point then is that the objective decision maker isn't just one kind among others, but he is the kind, the only kind, whose decisions have a peculiar force when communicated to others. His decisions put others in a bind. If one says, do this, I would like that very much, we reply easily, good, good, but not I. But if one says, do this, for such and such relevant reasons, we may still reply, but *I* prefer not to; but now our position is much weaker in an important respect, since he has said something that has force for us as well as himself, and we have countered with an alternative that has force if at all only for ourselves. We have not *replied*, if that means countering with an alternative we can seriously recommend to *him*.

These remarks support the observation that when one raises a moral question, wonders what he ought to do, thereby expressing moral puzzlement, he is asking what an objective decision maker would do. For in any case he is asking what would be right to do as such —objectively— and this character of objectivity in a decision is tied to objectivity in the decision procedure, in the sense that it is only when our decisions coincide with those an objective decision maker would make that our decisions can be compelling as such and for all. Objectivity in the judgment follows upon objectivity in the judging, and judgments that purport to be moral purport to be objective.

We have arrived at the first station: One ought to be objective. As an in-station entertainment, it may prove interesting to consider whether the proposition that one ought to be objective has implications for current ethical points of view. Existentialism suggests itself in this connection, because it tends to glorify subjectivity. The relevant implication seems to be that existentialism is fundamentally wrong-headed, in the same way that anyone who consistently gives wrong advice is wrong-headed. This conclusion is unavoidable if it is accepted that the distinctive feature of existentialism is endorsement of the idea that when one is forming an opinion the matter of overriding importance is the relation of the opinion with oneself as its believer. Truth in the important sense is held to be a property of that relation, or a consequence of it, and *what* one opines should be regulated by the requirement that it should be true, in the sense of true to oneself. It is a question of fundamental orientation, of the direction in which to look when troubled or in doubt. The idea of objectivity is that one is to look wherever the relevant factors are, that these factors are likely to be anywhere, and that for the most part they will not be in or parts of the needs and experiences of the one troubled or in doubt. Consequently objectivity obliges us to believe what the situation demands of us, regardless of our ability to incorporate the belief. It is entirely open to one who endorses the idea of objectivity to favor a relation with beliefs in terms of which the believer believes deeply, and is prepared to stake his all on it; what is not open to him is that his believing should be directed by that requirement. But the doctrines of subjective truth and the subjective thinker, and existentialist variations on that theme, appear to imply that believing should be so directed. Otherwise the subjective thinker would be "untrue" in his own sense to his beliefs, since he would be accepting a method of fixing belief that is unrelated to his own peculiar notion of wherein the truth of beliefs consists, and in this way would be showing his lack of commitment to that notion.

The offensive reply to this consists in reiterating the commitment to subjective truth, and defying the one who prizes objectivity to find the point of view from which the commitment may be tellingly criticized. But this is easily done. For it has been shown that we ought to be objective, and from this it follows, to put it broadly, that we ought not to be existentialists —in case, that is, being an existentialist means being guided in one's belief fixing activities by the ideal of subjective truth or one of its variants. (There is a tacit distinction here between *being* an existentialist and talking existentialism or writing about it in an approving way.) When one is told (truly) and accepts, that he ought not to be what he is, that is an end of *discussion.* He may reply, so be it, but I'll be it, and of course he may —the situation is quite like that of one who denies the law of non-contradiction; he can do it if he can do it, but if he does, then some things become true of his position —so far as he lives it— is that for better or worse, as it were, it is immoral, if being immoral means doing as one oughtn't.

I turn next to the question, Is objectivity a merely formal principle, or is it substantive? Knowing that we ought to be objective, are we then enabled to know some of the actions we ought to perform? Or is it as some have said of the categorical imperative, or of the part of it that can be defended, that though it may have some interest it has no use, since by its means we may never discover what we are to do. Now, there was an additional point to the instation entertainment. The discussion of existentialism established that the principle of objectivity is substantive. Objectivity requires that we not be existentialists, in the sense attributed to that position, and this is a substantive directive, as far as it goes. It demonstrates, that is, that knowing that we ought to be objective we are enabled to know at least one action that we ought to perform —if it is intelligible to speak of the act of not doing so and so; otherwise, what we know is a class of actions, that class marked by the characteristic of having been chosen because they seem to meet a requirement that makes no special reference to the sentiments of the chooser. And in any case this shows that the norm, Be objective!, is substantive, since it marks off the class of actions it endorses as a sub-class within a wider class, the rest of which are excluded by the norm.

We have, then, unexpectedly accomplished the second leg of the journey, none the worse, I hope, for wear. We get at least some direction from the norm, Be objective!, although admittedly that direction is not very definite. Before beginning the third leg of the journey it will be well to establish the importance of the minimal conclusion reached. There is no combination of plausible moral principles that make it not the case that some actions are arbitrary in the sense that appeal to those principles will not enable one to know of those actions whether they ought to be carried out. That is, we can never find a non-arbitrary decision procedure that in principle would enable one to discover of every possible action whether it is preferable to its alternatives. Of some actions it can neither be said that they ought nor that they ought not to be carried out, and there will never be a time when a more accurate understanding of morality has been attained, in terms of which those actions that now seem to be arbitrary are found to be non-arbitrary.

It would appear, then, that there are two classes of actions: those that one either ought or ought not to carry out, and those of which this cannot be said. Call them actions of the first, and of the second class. Actions of the second class are all right; that is, just, they are not wrong. But it follows from the fact that both classes have members that the apparent limitation of the case for objectivity —that the norm, Be objective!, is not very definite in the sense that there are plenty of alternatives which it inclines us neither to accept nor reject —is less serious than it might otherwise seem, since no combination of plausible norms will be entirely definite. The defect, if it is a defect, is at most one of degree; the patient may be ill, but he isn't dead —yet.

The last leg of the journey is directed toward the proposition that we can build the whole notion of morality on the premise, Be objective! I want to argue that the entire range of actions that in fact ought to be preferred —actions of the first class (ignore for convenience those in that class that ought *not* to be preferred)— would be preferred by an objective decision maker who assumed no norm or moral principle other than that he ought to be objective. The point of the last in-station entertainment was that it prepares one for not rejecting the position to be defended on the ground that the bare requirement —Be objective!— does not enable one to decide every issue. It does not permit this because not every issue *is* decidable; but it alone —that is, apart from all other norms or moral principles— permits one to decide those issues that are decidable.

I take advantage of the idea that it is possible to derive second order norms from the norm to be objective. Consequently, if one would hold that some decisions are correct because of the outcome of applying, say, the principle of utility as a standard, to the alternatives, I could consistently accept that provided that the principle of utility is such a second order noun —that is, provided that it gets its plausibility from its derivability from the norm to be objective.

The argument here has two parts. First, I want to indicate how second order norms may be derived by actually deriving one. The assumption I make is that many other norms are similarly derivable, and if they were derived the conceptual apparatus would be at hand for making good on the claim made —that objectivity is an adequate base for the whole of morality. Second, I want to develop a completely general argument for that claim, an argument that shows that unless a second order norm is derivable from the norm to be objective, it has no claim on us. The two lines of argument converge. The first assures that there are actions of the first class (in case the earlier argument to the effect that we ought not to be existentialists seemed tenuous). And the second line of argument assures that if there are actions of the first class, then they are such because they would be preferred by an objective decision maker who assumed no norm other than the norm to be objective. The result is that if one becomes disturbed by the number of actions of the second class —thinking that the requirement to be objective will not permit one to justify, say, all of Ross's prima facie duties —the prima facie duty of promise keeping, for example— we may expect him to rest content with the reflection that he (and Ross) were wrong, that some of the kinds of action they thought binding on us really aren't, and we may expect him to take heart by reflecting that now he has a point of view from which to defend his moral convictions, so far as they are to be defended, and need not fall back on the "defense" of intuitionist-rationalists, that it must be so because I (and other right and good thinking persons) can't help thinking so.

Discussion of the first part of the argument must rest on an account of what it means to "derive a second order norm from the norm to be objective." We may speak of a norm having force. It has force in case it is not arbitrary; that is, in case we ought to follow it. This is the same as its being a norm. If it has no force it is *only* a putative norm. But a norm may not have force in every context. It is a question of whether the context is such as to make the norm applicable. The norm not to murder has no force in a context where malice cannot appear; the norm not to steal has no force in a context where there are no property rights. There is even a context in which the norm to be objective has no force —that in which information is so uncertain that a mechanical or chance procedure for selecting among alternatives is preferable. A norm ought to be acknowledged in any context where it has force. A putative norm is a directive that purports to have force: what it purports is that it ought to be acknowledged in those contexts where its meaning is such that it would have force if it were a norm. Acknowledging a norm means allowing one's decision on an issue to be affected by the bearing on the issue of the norm. It can't mean deciding the issue in the way the norm suggests, since more than one norm may have force in the context where the issue lies, and another norm that has force there may suggest a different decision.

So far, the only norm identified is that we ought to be objective(O). "Deriving" another norm from this one consists in showing of some putative norm (pn) that if it is not acknowledged (pn) in a context where it would have force were it a norm, then the norm to be objective is not acknowledged in that context (O). Consider only those contexts where O has force. If a putative norm is derivable from O, the putative norm is a norm. This for the following reason. A putative norm's being derivable from O means that not acknowledging the putative norm is a sign of failing to acknowledge O. But this means that acknowledging the putative norm is a necessary condition of being objective. We ought to do anything that is a necessary condition of doing what we ought to do. We ought to be objective. And so it follows that, if a putative norm is derivable from O we ought to acknowledge the putative norm. In that case, the putative norm *is* a norm, since what it purports is that it ought to be acknowledged in contexts where it has force.

Whether a putative norm is derivable from the norm to be objective is decidable by asking whether failure to acknowledge the putative norm signalizes failure to be objective. In asking this one is considering whether a certain kind of implication (Q—), analogous to formal implication, holds between the fact of acknowledging (or failing to acknowledge) one norm, and the possibility of acknowledging another. Discovery that failure to acknowledge one norm implies not acknowledging another is the same as discovering that acknowledgement of the second is impossible when the first *is* not acknowledged. (a "derivation" is a proof that (pnQ—O).)

Here is a derivation. The putative norm to be derived is the duty of fair play: if some people are living together under a system of coercive rules, and by following the rules contribute to a condition from which all in the group benefit, then each member of the group, so far as his enjoyment of the benefit thus created is voluntary (he could leave the group if he wished to), is under a duty of fair play to abide by those rules. Thus, I have a duty to obey the traffic laws in my community because I benefit from the fact that others obey them; if I disobey, I withhold from the others a benefit that I voluntarily assume for myself, and this violates the duty of fair play. The derivation consists in the following considerations: If one does not follow the rules, obedience to which by others benefits him, he is choosing for himself fewer restrictions or greater freedom than is allowable to the others (if the benefit is to be realized). In so doing (failing to acknowledge the norm, play fair) he is failing to acknowledge O, since he is assigning more weight to his own case than he is to that of the others. He is being partial, and failing of objectivity, by treating two relevantly similar cases in relevantly dissimilar ways.

Since we cannot violate the duty of fair play in a context where it has force and be objective in that context, it follows that in such contexts we ought to acknowledge the duty of fair play.

The second part of the argument is that there are no norms that aren't derivable from O. This is the converse of the claim made above, that *if* a putative norm is derivable from O, then it is a norm. ($[\mathrm{pn} \, {-\!Q\!-} \, \mathrm{O}] - \mathrm{pn}$ is no norm.) That is, if acknowledging a putative norm is not a necessary condition of acknowledging O, then it is not the case that the putative norm is a norm. If this be true (if it holds for every putative norm) then O is a complete base for morality in the sense that everything that we ought to do ought to be done because failure to do it signalizes failure to be objective. Or, our whole duty is summed up in (and in whatever follows from) O.

The basis for this position was laid down in the first part of the paper, where it was argued that doing as we ought and doing what one who is objective would do come to the same thing. Owing to this, we can replace O with the directive, "Do as you ought!" Consequently, (1) since deriving a norm involves discovering that failure to acknowledge it signalizes failure to be objective, it also involves discovering that it signalizes doing as one oughtn't. But (2), owing to the identification of "ought" with "objective", there is no way of failing to do as one ought other than by failing to be objective. Therefore, if it is not the case that not acknowleding some putative norm implies not being objective ($-[\mathrm{pn} \, {-\!Q\!-} \, \mathrm{O}]$), then it is never the case that one who fails to acknowledge the putative norm does as one oughtn't. And this means that the putative norm is no norm—for a putative norm's being a norm consists in its being the case that there is a context in which it ought to be acknowledged.

Again, what assures us that the underivability of a putative norm implies its not being a norm is that (1) its being a norm implies that there is a context in which we ought to acknowledge it, and (2) saying that we ought to acknowledge it is a way of saying that an objective decision maker would. But the underivability of the norm means that an objective decision maker (qua objective decision maker) wouldn't acknowledge it. All that an objective decision maker, qua objective decision maker, does is follow the requirements of his status; that is, acknowledge all and only those norms acknowledgement of which is a necessary condition of his remaining objective. Whatever putative norm is not such a necessary condition (=is not derivable) is not binding (=is no norm).

II

We have arrived at the last station. If the argument is sound, some important conclusions concerning morality have been reached: that there is such a thing, that it is not a matter of opinion or taste and that when we say of some action that it ought to be performed, this is either true or false. And we now know why it is true: it is true because it follows from the norms that have force in the context in which the issue lies. The norms that have force there are norms because and only in case they are derivable from O. And O is a norm because it explicates the meaning of "ought".

The final in-station entertainment occurs in two acts. The first concerns the relation between conflicting norms. The second is highly speculative. 1. What are we to do in case two norms have force in the same context but suggest different decisions? Suppose we can derive both that an act is to be preferred so far as it spreads benefits evenly, and that it is to be preferred so far as it maximizes benefits. Then, to adapt a familiar example, we might be hard put to know what to do if in some context there are two ways of bringing some benefit to a community —one of which would provide a maximum amount of that benefit, the other of which would more fairly distribute a lesser amount. It is possible that another norm is derivable that permits a ranking of these two, or that suggests a manner of weighting the values they identify —quantity of benefit and the fairness of its distribution. But suppose that there is no such norm.

The problem that arises is that no course of action in the situation is binding on us unless it satisfies the requirement of objectivity better than its alternatives do, and to all appearances there is no way of settling the conflict by appealing to that requirement. But we think that most problems are like this: that they involve a conflict of plausible norms, that what we ought to do cannot be known unless it is known which of the conflicting norms ought to be ranked highest, and that the bare idea of objectivity doesn't provide or suggest a ranking of the norms. And so it would seem to follow that in most cases though there are many demonstrably wrong things to do, there is no one correct thing to do. If this is so, then we must of course accept it. It would not jeopardize the position defended here. But we ought to consider closely whether it is so.

I suggest the following way out. One who is reaching for objectivity *may* not find it possible to discriminate among the only available alternatives by appealing to norms or to the weights of conflicting norms. But he may nevertheless have a recourse that is demonstrably superior to tossing a coin. He may simply entertain the alternatives, independently of norms but, so far as he can attain it, in his status as objective or disinterested decision maker, and from this perspective he may nevertheless prefer one of the alternatives. When this is the case, the preferred alternative ought to be accepted. It ought to be accepted because it is preferred by an objective decision maker, and, as we have seen, what objective decision makers prefer is what ought to be preferred. Whether the person *is* objective, or disinterested, at the moment of discovering in himself a preference is of course a very important question and one not easy to answer, particularly by the person himself. But *if* he is, then we cannot doubt that he has found the right course of action.

The difficulty of knowing whether one has considered a matter disinterestedly should not be overstressed however, Often when we have evolved a preference in a matter we can be reasonably certain that we have done so from a standpoint of disinterestedness. We can discriminate among problems we confront and make fairly reliable estimates of the differing degrees of disinterestedness we are likely to bring to them, and we can often think of ways of safeguarding ourselves against a tendency to be partial in a matter. Partiality, subjectivity, interestedness, are not amorphous tendencies that we express in an undifferentiated way in any and all situations; they are instead provoked by specific and identifiable features of situations, and making the identification is often a useful step in overcoming the tendency. Moreover, we have moderately well-founded views concerning the ability of our friends and associates to decide matters of various sorts objectively. If we don't trust ourselves to make an impartial settlement of some issue, and seek advice from another, there are some we wouldn't think of approaching, since we are reasonably certain that they are as subjective as we are on matters of that kind; and there are others on whom we would reasonably bestow considerable confidence. And if we are asked to justify these opinions we hold about the capability of ourselves and others to be objective in considering issues of various kinds, there are relevant and plausible things we can say —that such and such a person has a very bad record on issues of such and such a kind; that he cares too deeply about some facet of the issue to reach a balanced view; that he is constitutionally unable to grasp the purport of all the relevant facts; that he lacks the required information, or the required sensitivity; that the situation threatens his self-esteem too much for him to judge it accurately —or his pocket book, or his life. And so on.

The second-mentioned manner of deciding what one ought to do —by accepting what one finds oneself preferring while contemplating the whole issue from a standpoint of disinterestedness— suggests the second way in which Kant would have us apply the categorical imperative. In the case where the maxim of the act is capable of standing as a universal law, we may nevertheless discover that as rational creatures we cannot will that it should so stand. The similarity between Kant's position and the one taken here results from the similarity between appealing to what an objective person would prefer (will), and appealing to what a rational person would prefer. My inclination is to suppose that in the context "rational" means disinterested or objective, in which case the "similarity" is rather an identity.

What if it is not possible to decide what to do by appealing to norms derivable from O, nor by consulting one's preferences as a disinterested decision maker? Clearly there are such issues. But though they may be vexing to those who try to decide them, they create no moral problem, and no problem for ethical theory. The reason is that in such cases there is no alternative that one ought to prefer. Since the belief that we ought to do something *is* the belief that that is what an objective or disinterested decision maker would decide to do, issues that a disinterested decision maker can't decide are issues in which none of the alternatives "ought" to be done. It isn't only that in those circumstances we don't *know* what ought to be done; rather, we know that *none* ought to be done.

2. By way of conclusion now I want to just barely point at some of the further implications of the idea of objectivity. Suppose a person were thoroughly objective —that means he would be utterly moral, not merely in

what he does but also in his feelings and inclinations. Then in what sense would he be an "individual"? What use would he have for the word "I"? In what sense would he be "self"-conscious? I think that the answers to these questions are so strange that we must suppose that a thoroughly objective person would be radically different —different in kind rather than degree— from the person we meet in the mirror and in the world around us. So different that if in our thinking about the problems of personal identity, the nature of the individual, the meaning of freedom, and the relation between consciousness and the world, we had in view that kind of person as the subject of the attributes we are investigating —*his* freedom, consciousness, personal identity, and individuality— we would get radically different answers from those we ordinarily get.

My reason for saying this is that I think a person who never desired anything *for himself* would not apprehend himself as a subject, and therefore would not use the first person singular pronoun, or at any rate would not use it in the way we do. If all of his desires were referred to an object of which he considered himself a part —so that any condition he desired for his own bodily frame, say, would be desired for the sake of that larger object, as contributing to a desirable condition of *it* —then for him the subject of what we would call *his* actions and thoughts would be that object. But an utterly objective person *may* find that the human race is such an object, and we may find that he perceives all of "his" actions and thoughts as its, so that when he says "I" —meaning thereby the subject of those actions and thoughts— he is referring to the race.

This is no idle speculation. One ought to be objective, and therefore there are imaginable circumstances in which the fact one conceives oneself in the way we normally do would be a sign of immorality, of one's not being as one ought. I imagine that the indicated circumstances are those in which one has reason to think that all others will be objective. That is, when I have reason to think that "my interests" will not be fairly represented in the decisions of others, the objective course for me to take involves identifying myself as a point of reference for desires, hence adopting or becoming the normal "I". But as soon as I have reason to think that others will treat me with complete impartiality, for me to form desires for myself —to have and pursue them qua *mine*, as distinguished from identifying them as desires the race (in me) should have satisfied— is for me to become immoral.

The result seems to be this. If we feel, as I suppose we are bound to feel, that the loss of that plurality which results from everyone referring by means of "I" to a different subject, would be unfortunate, then we should be delighted that not everyone is moral.

Now you can get off.

University of Waterloo,
Waterloo, Ontario.

Aristotle and Virtuous Activity

by Richard P. Haynes

Many of the difficulties that one encounters in Plato's discussions of *Arete* are due to the fact that the word *arete* has both the very general sense of "excellence" or "goodness" and a secondary sense roughly equivalent to the English word "virtue". We even use "virtue" in this broader sense, as when we ask, "What is the virtue of doing that? " or say, "He has all the virtues of a good general." If the word *arete* has one common meaning, it is something like "what we mean in calling something good of its kind". Difficulty is bound to occur as soon as we ponder the reasons why we call each of the following items good: an eye, a brick-layer, a governor, a man. *Arete* is not what makes a knife a good thing, but rather what entitles us to call a particular knife a good knife. Plato frequently treats the character traits we call virtues as though they were similar to other forms of goodness, talking in the same breath about being versatile at cobbling, horse-training, and justice, and it is never really clear that Plato managed to see the dissimilarities between the virtues and the forms of technical goodness.(1)

Von Wright thinks that Aristotle too was misled into thinking that virtues and abilities or skills are more closely similar than they actually are.(2) There are at least two very good reasons for thinking that von Wright's suggestion is correct.

1) The analogies that Aristotle draws in Book I of the *Nicomachean Ethics* between the criteria of the goodness of a good man and the goodness of a good shoe-maker, harp-player, or eye lead us to believe that just as a good shoe-maker is called good because he is *good at* doing something, so a good man is called good because he too is *good at* doing something. And since "good man" and "virtuous man" mean the same thing, virtue is a kind of *technical goodness*, an ability to do certain kinds of things well.(3)

2) Even though Aristotle seems to understand the way that the virtues do in fact contribute to human welfare, he also quite clearly wants to think of engaging in virtuous activity as an important kind of skillful, and hence enjoyable, activity, done for its own sake and contributing immensely to happiness.

In what follows I shall examine these two points in order to determine why Aristotle was led to think of the virtues as forms of *technical goodness.* We shall find that Aristotle gives a biological sense to the notion of *human excellence*, and that this is the source of part of the difficulty. He also equates *eupraxia* (prosperity) with acting virtuously, and *euzoia* (living-well) with doing something skillfully. We may state the source of the trouble in this way. Aristotle conceives of living happily as engaging in activities for their own sake, for if everything we did we did in order to obtain some results, the ideal life would be a life of inactivity, but if a life of inactivity is a happy life, we might as well "call stones and corpses happy," Callicles complains,(4) and this is a just complaint against some of Plato's more morbid inclinations.

Hence, there must be activities that are worth doing in their own right and for no end outside the activity itself. Such activities are done merely for the pleasure of it, and the highest of such activities constitute the happy life. But now Aristotle wants to equate activities done for their own sake with activities done with human excellence, and for several reasons. 1) Skill contributes to the enjoyment of an activity, in fact, enjoyment simply is unimpeded activity; 2) the human excellences that we call virtues (the excellences of character) are a) done for their own sake, and b) defined as contributing to human welfare; and 3) living well or doing well (for oneself) must mean doing something skillfully, so that a happy man is a good man, and a good man is obviously a man of good character, therefore, virtue must be a skill in engaging in uniquely human activities.

I. *Human Excellence.*

The distinction between virtue as a character trait and the other kinds of *arete* is often masked by social classifications. This is illustrated in the *Meno* in the context of a discussion whether *arete* comes by nature, practice or schooling. It is clear from that context that they are concerned with that sense of *arete* that the Sophists offer to teach and that makes one a good man —that is, a man looked up to by others. "What is *arete*? " Socrates asks Meno. Meno replies,

> A man's *arete* is that he be competent to manage the affairs of the city, and to manage them so as to benefit his friends and harm his enemies, and to take care to avoid suffering harm himself. The *arete* of a woman can be described as ordering the house well, looking after the property indoors, and obeying her husband. And the child has another *arete* —one for the male and one for the female, and there is another for elderly men, and one for the freeman and one for the slave....(5)

Here the word *arete* holds a half-way position between the general sense of "excellence" and the special sense of "virtue". To talk about good men or women, good husbands or wives, good fathers, or good servants is not entirely analogous to talking either about a good person or about a good brick-layer. "Husband," "wife," "servant," "governor," even, in most cultures, "man" and "woman" are office-concepts. A man is a good brick-layer if he has mastered a certain skill, that is, whether he can do certain things well, whether or not he ever does them in the future. One can have skills and choose not to use them, so a person can be a good bricklayer without actually *being* a brick-layer. However, one cannot be a good father without actually *being* a father. If someone were not a father, we might say that he *would make* a good father, but not that he *was* a good father. Being a good father, therefore, is not simply *being able* to do certain things well, but it involves doing them and doing them conscientiously. We could say that the idea of character is part of the excellence of office-concepts. Although Aristotle seems to get away from the implications of social classification when he talks about human excellence,for he interprets "human" as a biological concept, one suspects that a good deal of what Aristotle thinks of as virtue consists in engaging in gentlemanly activities. The idea of human excellence in Aristotle is a fairly confused one. On the one hand, he carries out the biological concept of excellence, confusing it to some extent with forms of technical goodness, and on the other hand he adds the socially commendable qualities whether they fit the biological concept or not.

> Every excellence makes the thing that has it a good thing and also allows it to perform its function well. For example, the effect of excellence in the eye is that the eye is a good one and functions well; since having good eyes means having good sight. Similarly, excellence in a horse makes it a good horse, and also good at galloping, at carrying its rider and at facing the enemy. If this is true of all things, excellence in a man will be that condition which makes him a good man and also which will cause him to perform his function well.(6)

Taking the biological point of view, Aristotle distinguishes three forms of human life. The human *psyche* has 1) an irrational part that is obedient to reason, 2) a rational part that deliberates and 3) a rational part that engages in theoretical activity. The excellences of the irrational parts are temperance and courage,(7) of the rational part that deliberates, wisdom *(phronesis)*, and of the rational part that theorizes, scientific sagacity *(sophia)*. The distinction between excellences of character and excellences of intellect runs parallel to the above distinctions, but it is slightly different from it. Any acquired condition that is worthy of praise, Aristotle says,(8) is an excellence, and we praise a man for his intellectual qualities as well as for the qualities of his character. On this basis we discover a larger number of excellences, including the other-regarding virtues and the social amenities. *Techne* is rejected as a human excellence for it is concerned with making rather than with good behavior. Although there is such a thing as *technical excellence* (excellence in making), this excellence is not a human excellence, for it does not entitle us to call its possessor a good man, but only a good so and so, where "so and so" is the name of some craft. Nor is a bad doctor necessarily a bad man.(9) However, both the human excellences and the technical excellences belong to the same genus, that of *hexis.*

A *hexis* is like a natural capacity in that we can say that a man is a certain kind of man because he possesses a certain one.(10) But unlike the natural capacities it is acquired.(11) We do not call a man good nor do we praise him for his natural equipment.(12) At least some of the acquired conditions are essentially related to natural capacities. There is the natural capacity to feel fear (i.e., to be affected or moved by the presence of danger), and then there is the acquired condition of being able to feel fear in a certain way. The courageous man feels fear only at the right time and to the right degree. Courage seems to be an acquired capacity to feel fear. An acquired condition is differently related to activity than a capacity is, for we first have the natural capacity, and then exercise it. But acquired capacities are acquired by engaging in those acts that define the acquired capacity. One becomes good at playing tennis by playing tennis, and playing it well. Similarly, one becomes courageous by doing courageous acts. But there is an important difference between what defines a virtuous act and what defines a skillful act.

II. *Virtuous Acts.*

The acts connected with virtue and those connected with *techne* are essentially different, for *techne* is concerned with making, and making is bringing something into existence. The product may be judged on its own merit, and even though a product may occasionally come into existence by chance, the excellence of the product is generally the criterion for the excellence of the *hexis* from which the product came. But there are no products of virtuous acts, so we must judge them by the condition from which they arise. Here Aristotle is not as clear about difference as he should be. We should distinguish, as von Wright does, between acts denominated by the results produced, activities which are the exercise of inward states (abilities) but which have very definite outward criteria, and those acts which derive their name solely from the inward state that gives rise to them. Building a house, opening a can, driving a nail, are acts named after their results. One feature of these acts is that if one stops doing them before a certain result is achieved, the acts remain incomplete, but when the result is achieved, the act must cease. Running, whistling, dancing, are activities, and they are complete at any stage in their exercise. Whistling a certain tune or dancing a certain dance are performances that have features of both acts and activities. They are activities in that they may leave no result behind them, but they are like acts in that they remain incomplete until a certain stage has been reached in their performance. Virtuous acts are neither acts nor activities, since an act of courage produces no characteristic result, nor does it have any characteristic outward criteria. Acts of courage are acts that we would expect from courageous people under certain circumstances. But if you tell me that John is in the next room performing an act of courage, I have little idea of what he is actually doing. One and the same act or activity can, in one circumstance, be an act of courage, and in a different circumstance, an act of cowardice.

Although Aristotle does distinguish between activities and productions (acts), he seems to confuse this distinction with another one, namely, the different reasons for which someone may do something. He seems to suppose that since producing is bringing some product into existence, the reason that one produces is because one wants the product. On the other hand, since there are no typical products of activities, the reason that one engages in them must be because they are worth doing in their own right. Since, Aristotle says, the quality of a virtuous act is not determined by the results it produces, it must be an activity.

Aristotle is very clear about the value of being virtuous. Since our emotions are natural spurs to action, in a rational individual they will be trained to follow reason one way or another. The natural capacity to feel fear is replaced in the courageous individual by a learned way of feeling fear, namely, at the right time, to the right degree, and so on. The courageous man has learned to make rational choices in fearful situations without suffering from the obstructing influence of fear. The temperate man is not overcome by desire for bodily pleasures, which means that his desire for these is not so strong that it obscures his judgment about the effects of indulging a particular desire on a particular occasion. The virtuous man has learned to be guided by principle rather than feeling, and he has learned this by using feeling to check feeling. The virtues are essentially beneficial, some to the individual who possesses them, some to those to whom the virtuous man relates. The *act of virtue* is precisely that act which the trainee must engage in to gain virtue, and although Aristotle seems to realize this, he also talks about the virtuous man displaying or exercising his virtue for its own sake, and this by doing virtuous acts.

In several passages Aristotle reiterates the point that the aim of the practical part of the *psyche* is *eupraxia*, which usually means something like "welfare," but which Aristotle takes to mean "good conduct." He also says in a number of passages that what characterizes virtuous acts is that they are done for their own sake, or, again, that all virtuous acts aim at *to kalon*, what is honorable.(13) I find this to be a puzzling point, and it is clearly connected with his claim that the test of virtue is not whether a person does certain acts, but whether he does them with pleasure, or at least without pain.(14) We may grant that a man has his desires for bodily pleasures mastered only when he no longer finds it painful to deny every little whim. But that does not mean that he needs to enjoy denying such whims. Does Aristotle mean that we are to take pleasure in passing up fourth helpings, or does he simply mean that we find being temperate enjoyable because when we are not tempted to over-indulge, we can live a pleasanter life? It is not clear. He does retract the suggestion that a courageous man must enjoy facing danger. The only qualification is that he must not be pained by it.

I suspect that Aristotle has put together several things and confused them when he says that virtuous action is done for its own sake and aims at what is honourable. These are 1) virtuous action is often acting on principle, and 2) not all virtues are directly beneficial to their practitioner, since there are both self-regarding and other-regarding virtues. These two points mean that not every virtuous choice is dictated by the results which would follow from some other choice, or which actually follow from this choice, and that not every virtuous act promotes the welfare of the agent. But neither of these two points mean that virtue is not essentially followed because of what is beneficial I think it is because Aristotle is already disposed to believe that enlightened self-interest supports the practice of the other-regarding virtues that he is led to show just how it does this by treating all virtues as forms of technical goodness, the practice of which is essentially enjoyable.

This has particularly odd results when applied to some of the virtues. It is quite one thing to think of an altruistic individual taking pleasure in helping people because he is distressed by the suffering of others or because he likes to see others made happy, and quite another to think of the same individual glowing with pride at the display of his beneficence as he hands over a check to his favorite charity. Or again, to think of someone taking great pride in his display of justice as he keeps a contract by paying a debt, or returns a purse to an owner, or refrains from stealing some unguarded prize. While Aristotle seems to realize that there are certain difficulties to be encountered in trying to practice these virtues, this does not stop him from asserting that their practice constitutes a brand of happiness. And while he does say that it is more a quality of virtue to *do* noble acts than to *refrain* from shameful ones,(15) in Book X(16) he seems to ignore that distinction. While arguing for the superiority of activity of the highest excellence on the grounds that it requires fewer external goods, he notes that practicing the virtues often requires considerable wealth. At least the liberal man will need wealth to do liberal acts, and the just man will need the means to discharge his debts, and the courageous man will need strength to display his courage, and the temperate man will need opportunity for indulgence in order to resist it.

While it is true that certain kinds of acts taken individually can be classified as virtuous acts, it is not true that we can think of virtue as a kind of skill that one practices from time to time. The virtuous man is not virtuous because he is good at doing virtuous acts. We call an act a virtuous one if either a) it obviously shows that the agent is virtuous because he has done something that few people are able to do, like risk one's life for a good reason, or b) it is the kind of thing that people who are training themselves to be virtuous would do when it is obviously hard for them to do it. But it seems foolish to think of the virtuous man running around looking for opportunities to display or practice his virtues. Should we expect the temperate man to seek out temptations to resist, or the just man to enter contracts so that he may discharge them, or the courageous man to seek out danger in order to face it bravely, or the man of gentle temper to seek provocation? Such behavior would more properly characterize someone who is trying to prove himself.

It seems as though Aristotle did assimilate the virtues to the other kinds of excellences, those involving technical goodness. We have seen why he might have been led to do this. Because a) he thought of virtuous acts as different from productive acts, the difference lying in the fact that the virtuous act derives its name from the character of the agent, while productive acts derive their name from what is produced, and b) since we engage in productive acts because we want the products, we must engage in virtuous acts for their own sake, since there are no relevant products. And since c) we enjoy exercising our skills and often do so because it is enjoyable, the reason that we engage in virtuous acts is because it is enjoyable and hence is an exercising of skill. And all of this conforms to the fact that just as the good general is good because he has certain technical excellence, i.e., is good at doing certain things, so the good man will also be good because he is good at doing certain things.

NOTES

1. On this point see John Gould, *The Development of Plato's Ethics* (Cambridge, 1955), Chapter II.

2. Georg Henrik von Wright, *The Varieties of Goodness* (Routledge and Kegan Paul, 1963), p. 137. I am much indebted to von Wright's excellent book, for I have followed out many of his suggestions, and I have relied heavily on the many distinctions he makes.

3. The expression "technical goodness" is not to be confused with Aristotle's use of "*techne*" as essentially a *poiesis* or making.

4. *Gorgias* 492 E.

5. 71e-72a.

6. *E.N.*, 1106a14-25.

7. 1117b23.

8. 1103a8-11.

9. 1148b8.

10. At 1105b19 Aristotle distinguishes three types of psychological concepts, *hexis*, *pathos*, and natural capacity. These roughly parallel the first three kinds of quality given in the *Categories*, 8b25-10a10, except that a *pathos* is not a quality.

11. Although Aristotle uses the expression "the natural hexeis" at 1144b9.

12. 1106a7-13.

13. 1117b12, 1122b7, 1176b6, 1104b30.

14. 1104b5.

15. 1120a10.

16. 1178a30.

University of Florida,
Gainesville, Florida.

The Birth of Pragmatism:
Peirce's Attempt to Solve the Problem of Induction

by Donald Koehn

I

In 1877-78 Peirce published a series of six papers in *The Popular Science Monthly* entitled "Illustrations of the Logic of Science."(1) Since then the texts of all six papers or edited versions of them have appeared in three different collections of the writings of Peirce.(2) The first two papers, erroneously called "The Fixation of Belief," and "How to Make Our Ideas Clear" are now recognized as the first public statement of what came to be called "Pragmatism."(3) Edited texts of at least one of these two papers have been reproduced in at least fifteen different books.(4) Naturally the commentary on the first two is much more extensive than that on the others. Nevertheless, some discussion of all six papers has now been published, and a complete bibliography of all commentary on the six papers would contain at least one hundred items. In spite of the wide circulation of the papers and in spite of the large quantity of commentary, no paper, since 1877-78, has been published with the correct title, and none of the papers has been interpreted from the point of view from which they were originally published.

Peirce's title and running head for each paper is the same: "Illustrations of the Logic of Science." The subtitle of the first paper is "First Paper — Fixation of Belief." The subtitle of the second is "Second Paper — How to Make Our Ideas Clear." In the same way, the subtitles of each of the papers contains the phrase which marks the place of the paper in the series. Unlike any edited version of the various papers, the running head of the original is not part of the subtitle, for example, "The Fixation of Belief"; rather it is the same for any paper, "Illustrations of the Logic of Science." The fact that Peirce uses the same running head for each paper and that he gives a number with each subtitle suggests that he was interested in the series as a whole and that the individual papers were conceived as parts of a larger whole in which their sequence is important. Commentary has generally dealt with the various papers as though each was an individual essay upon an isolated topic. Some commentaries treat the first two papers as a single unit, but there is no commentary which considers the possible function of any of the papers in the whole series. If, contrary to the approach of existing commentaries, Peirce published the papers in the way suggested by his titles and running head, then an accurate and thorough analysis of any one paper would consider its role in the series as a whole. In this essay I shall argue that the six papers of the "Illustrations" were published as parts of a unified discourse on the "logic of science," and I shall show in what their unity consists. My purpose is not to analyze any one paper, or the series. I shall show the direction to be taken by an accurate, thorough analysis of the "Illustrations of the Logic of Science."

II

A rapid survey of all six papers is likely to support the view that Peirce published them as individual essays on disparate aspects of the logic of science. The fact that the term 'illustrations' appears in the title suggests a more or less random, disconnected treatment of the logic of science. The topics of the individual papers reinforce this suggestion. What does a discussion of the clarity of ideas (second paper) have to do with the order of nature (fifth paper)? Why should the latter be discussed just after a discussion of the probability of induction (fourth paper)? While the numbering of the papers and the fact that the running head of each is the same suggests that Peirce is trying to present a single unified treatment of the logic of science, other aspects of the titles suggest the contrary. Furthermore, if the texts of the papers are surveyed rapidly there seems to be no plan whatsoever for the series as a whole. I must admit that a first reading may very well leave a reader with the impression that Peirce took the opportunity of publishing a series of articles in order to present his views on a group of heterogeneous and disjointed topics. If the papers were published with a plan for the whole, it certainly is not obvious.

A careful study of the papers reveals passages where Peirce indicates what he takes to be the connection between some of the apparently disparate topics of the several papers. At the beginning of the second section of the second paper, Peirce says:

> The principles set forth in the first of these papers lead, at once, to a method of reaching a clearness of thought of a far higher grade than the "distinctness" of the logicians (12:289)

The method to which he refers is the rule for attaining the third grade of clarity (12:293).(5) The principles in the first paper to which he refers are: the sole object of inquiry is the production of belief, and belief is the establishment of a habit of action. While I do not think that the principles lead at once to the method, Peirce clearly tries to derive the method from these principles (12:289-293).(6) At the end of the second paper Peirce indicates that he is going to use this method in the third paper. He says that his reader will see in the next paper "how beautifully what has been developed in this tedious way can be applied to the ascertainment of the rules of scientific reasoning" (12:302). At the end of the second section of the third paper he says that certain difficulties in the calculation of probabilities can only be completely resolved by "making the idea of probability perfectly clear in our minds in the way set forth in our last paper" (12:607). In the third section Peirce immediately proceeds to clarify the idea of probability by use of the rule for attaining the third grade of clarity which was developed in the second paper, and in the fifth section he makes use of the clarification of probability to infer all "the fundamental rules for the calculation of chances" (12:615). Apparently, he takes this to be sufficient to settle the difficulties in the calculation of probabilities to which he alluded. If we take Peirce to be saying at the end of the second paper that he is going to use the rule for attaining the third grade of clarity to ascertain *all* the rules of scientific inference, then he is certainly not trying to do that in the third paper. Other rules of this kind are given later in the series. If we take him to be saying that in the third he will exhibit *some* rules of scientific method by the rule developed in the second, then he has fulfilled his promise.

At the end of the third paper Peirce says:

> According to the view here taken, these are the only fundamental rules for the calculation of chances. An additional one, derived from a different conception of probability, is given in some treatises, which if it be sound might be made the basis of a theory of reasoning. Being, as I believe it is, absolutely absurd, the consideration of it serves to bring us to the true theory; and it is for the sake of this discussion, which must be postponed to the next number, that I have brought the doctrine of chances to the reader's attention at this early stage of our studies of the logic of science (12:615).

A sketch of the argument of the fourth paper will show that Peirce brought the doctrine of probability to the reader's attention for the sake of that argument. He begins the fourth paper with a summary of the conception of probability and of the rules for its calculation which were delineated in the third paper. He then contrasts that conception (called the materialistic conception) with another (the conceptualistic). He then uses the rules for the calculation of probabilities to illustrate the difference of the two conceptions. This is done by first calculating a probability according to the rules he has given; he then gives an argument for and states the "additional" rule of the conceptualistic conception of probability; thirdly, he argues that we are led to an absurd result when we use the additional rule in the calculus of probabilities (12:707-11).(7) The comparison of the two conceptions is not left at this point. After distinguishing deductive and synthetic reasoning, Peirce argues that the conceptualistic treatment of probabilities is entirely inadequate to the representation of synthetic reasoning. In fact, Peirce argues, the answer to the problem of probabilities according to the conceptualists involves the assumption that the arrangements of Nature are equally probable. On that assumption we are led to the conclusion that "*synthetic* or loosely speaking *inductive* reasoning" (12:711) is "absolutely worthless" (12:714). Peirce then (12:714-16) turns to the materialistic conception and the problem posed by it. The solution to it is perfectly well known (12:715). After illustrating what can and can not be known about the probability of a synthetic inference, Peirce concludes the fourth paper with a brief section on the "paradox" (12:717) of synthetic reasoning. He states the "rule of induction" (12:718) and our reasons for knowing that it will hold good in the long run.

From this summary of the fourth paper we can see that Peirce attempts to show that synthetic reasoning can be explained by his conception of probability while it cannot be explained by another. The summary does not show just how Peirce explains the validity of synthetic inference nor how the explanation is based on his conception of probability. What is clear is that Peirce is trying to explain the validity of inductive reasoning by consideration of the two different views of probability. By tracing the implications of one conception Peirce is lead to conclude that synthetic reasoning cannot be explained; by considering the conception developed in the third paper, he is lead to show how it is explained.

At the beginning of the third section of the fifth paper Peirce says:

> In the last of these papers we examined the nature of inductive or synthetic reasoning. We found it to be a process of sampling (13:208).

He then argues that certain conditions must be met for the sampling process to constitute valid induction. His theory of induction is then contrasted to that of Mill, which bases the validity of induction on the uniformity of Nature. Near the end of the third section of the fifth paper, Peirce says:

> These are the essentials of induction, and they are not recognized in attributing the validity of induction to the uniformity of Nature. The explanation of induction by the doctrine of probabilities, given in the last of these papers, is not a mere metaphysical formula, but is one from which all the rules of synthetic reasoning can be deduced systematically and with mathematical cogency (13:211).

Peirce distinguishes deduction and the two kinds of synthetic inference, induction and hypothesis, in the first two sections of the sixth paper. Near the beginning of the third section he says:

> The same warnings that have been given against imagining that induction rests upon the uniformity of Nature might be repeated in regard to hypothesis. Here, as there, such a theory not only utterly fails to account for the validity of the inference, but it also gives rise to methods of conducting it which are absolutely vicious (13:475).

After he states the essential rules of hypothesis, he claims:

> The rules which have thus far been laid down for induction and hypothesis are such as are absolutely essential (13:477).

Other rules may be very valuable but the ones he states are essential for valid synthetic reasoning. Peirce apparently thinks that the explanation of synthetic reasoning in the fourth paper is one from which he has deduced the essential rules of both kinds of synthetic reasoning. The essential rules for induction are presented in the third section of the fifth paper; the essential rules of hypothesis are stated in the third section of the sixth.

I have given seven quotations which afford clues as to what Peirce himself took to be some of the connections between the discussions of the various papers. In the first five of these quotations Peirce explicitly refers forward or backward to another paper and he is explicit as to which paper he refers. In the last two quotations Peirce refers to the fifth paper,but he does not explicitly name the paper. By stating what Peirce does in some of the sections of the papers, I have shown that the clues afforded by the quotations reveal a continuous line of argumentation from the first through the sixth papers.

Principles in the first paper are used in the second to develop a method or rule for attaining a grade of clarity far higher than that given by other logicians. The rule developed in the second paper by principles in the first is used in the third to clarify the idea of probability and to deduce the fundamental rules of its calculation. The conception of probability and rules for its calculation developed in the third paper by the rule in the second are used in the fourth to explain the validity of synthetic reasoning and to show that it cannot be explained on a different conception of probability. The "theory of reasoning" (12:615) given in the fourth paper by the doctrine of probabilities expounded in the third is used in the fifth to infer the essential rules of induction. That theory is claimed to be superior to one which tries to explain induction by the uniformity of Nature. In the sixth paper the argument against basing induction on the uniformity of Nature is extended to hypothesis and the essential rules for its conduct are laid down. Unlike a theory of reasoning based on the principle of the uniformity of Nature, the theory in the fourth paper is one from which the essential rules of induction (fifth paper) and hypothesis (sixth paper) are deduced. We see then a continuous line of argument which begins in the first paper and runs through the last.(8)

III

We may now ask whether the part of the argument which is to be found in each of the papers constitutes the central point of the individual paper in which it occurs. If the several parts of the argument form a relatively minor or insignificant part of the individual papers considered in themselves, then the series might be a more or less random collection of articles on the logic of science with nothing more than minor points of connection between them. Because special qualifications will have to be made about the fifth and sixth papers, I shall consider them in the next part. Here I shall show in a summary way that the main points of the first four papers are parts of or are put forth for the sake of the argument which runs through the series.

The third, fourth, and fifth sections of the third paper are parts of the argument. The third and fourth sections are devoted to the clarification of the idea of probability and in the fifth Peirce presents the rules for its calculation. The first two sections are introductory, not only to the discussion of probability in the rest of the third paper but also to the theory of synthetic reasoning which is given in the fourth paper. In the first section Peirce claims that the history of science shows that the idea of *continuous quantity* has a great role to play in science, and that he intends to make great use of this idea in the series. He clearly tries to make use of this idea in his conception of probability.(9) In the second section he claims that the general problem of probabilities is simply the general problem of logic quantitatively conceived: How much are facts stated in premises worth as evidence to prove the possible fact in the conclusion of an argument? Peirce says that some believe that the paradox of the validity of synthetic reasoning can be reduced to a mere question of arithmetic. This belief is examined in the fourth paper.

He also says that the fundamental principles of the calculation of probabilities is in dispute and the dispute can only be settled by clarifying the idea of probability by the method laid out in the second paper. Thus, the second section serves as an introduction to two parts of the argument of the series. All of the third paper, then, is devoted to the argument. The first two sections are introductory to parts of it, and the main points of the paper in the last three sections, the clarification of probability and the statement of the basic rules of its calculation, are constitutive parts of it.

There are four sections in the second paper. The first is introductory. Peirce derives the rule for attaining the third grade of clarity in the second section. In the third he illustrates the use of the rule, and the fourth is devoted to clarifying the idea of reality. The only points which might be considered as major points of the paper are the development of the rule for attaining the third grade of clarity and the clarification of the idea of reality. The second section is part of the argument of the series. (We have already seen that Peirce uses the rule in the third paper to clarify the idea of probability.) We shall later see that the argument which runs through the series is more complicated than it now appears and we shall then see that the clarification of the idea of reality is part of the argument.

There are two points in the first section of the second paper that I shall want to consider later on. The first is that Peirce claims that the doctrine of clearness and distinctness of ideas is a striking example of how logicians have been slumbering during intellectual activity and never dreaming of applying the lessons of modern science to the improvement of logic. The second point is that Peirce claims that the *need* of attaining a higher grade of clarity is relative to the method of inquiry to be used. He claims that both Descartes and Leibnitz failed to develop methods of inquiry which required anything more than clearness and distinctness of apprehension. The *a priori* method requires nothing more. The third grade of clarity is needed by those who are to use the scientific method of inquiry.

Thus far, we have seen that Peirce used two principles given in the first paper to develop the rule for attaining the third grade of clarity in the second. I admit that the presentation of these principles is not the main point of the first paper. The main point is the analysis of the four methods of fixing belief. To show that this analysis is part of the argument which runs through the series I shall have to examine briefly the first paper.

In the first section Peirce implies that the subject of logic (as a study) is the art of reasoning and this is implicitly identified with the power of drawing inferences. After a brief historical review of different conceptions of reasoning and of scientific advances, he claims that each chief step in science has been a lesson in logic and an exemplification of the defective state of the art of reasoning when the step was taken. The last example he gives is the Darwinian controversy which he says is in large part a question of logic. Darwin proposed to apply the doctrine of probabilities to biology as Maxwell and Clausius had done earlier in the theory of gases. Just as these two could not specify the movements of any particular molecule, but could only predict what would happen in the long run to a body of gases, so Darwin could not say what the operation of natural selection and variation would be upon the individual organism but could only say that in the long run they would adapt animals to their circumstances.

In the second section Peirce infers the general criterion of the validity of reasoning from the end of reasoning. The question of validity is purely one of fact and not of our inclination to accept a conclusion from given premises. Our inferences are valid if the "guiding principles" (14:4) or rules of inference are such as generally to lead us from true premises to a true conclusion. Since almost any fact may serve as a guiding principle, a book cannot be written upon guiding principles unless it is limited in some way. There is a division of facts between all those which are assumed in asking the question of the validity of an inference and those not assumed in asking this question. For example, in raising the logical question it is assumed that there are such states of mind as doubt and belief, that a passage from one to the other is possible while the object of thought remains the same, and that this passage is subject to some rules which are binding upon all. He claims that it is easy to believe that "those rules of reasoning which are deduced from the very idea of the process are the ones which are the most essential" (12:5) and warns that it is easy to overlook the importance of what may be deduced from the assumptions involved in raising the logical question.

In the third section Peirce gives the three distinguishing characteristics of doubt and belief which are repeated in the second section of the second paper. In the fourth section he names any attempt to pass from doubt to belief "inquiry"; he argues that the only end of inquiry is to attain belief; and he claims that the latter is a very important proposition which sweeps away erroneous conceptions of proof. In the fifth section Peirce gives the now famous analysis of the four methods of fixing belief.

Normally commentary on this paper considers only the last three sections of the paper. It concentrates upon the doubt-belief theory of inquiry and on Peirce's discussion of the four kinds of inquiry. However, if the first two sections are read carefully it will be seen that Peirce's main concern in the first paper is not inquiry but logic, and the main concern of logic is valid inference and the main concern in this regard is to discover valid rules or guiding principles of inference. At the end of the second section it appears that Peirce, guided by lessons from the history of science, is going to give the essential rules of reasoning which can be deduced from the assumptions of the logical question. Peirce clearly does not tackle this job immediately. Rather he distinguishes doubt and belief which are facts presupposed in asking the logical question. He then infers the *genus* of reasoning from these facts. To ask whether an inference is valid is to assume that there are states of mind such as doubt and belief, and that a passage from one to the other is possible. Any attempt to make the passage from doubt to belief is an inquiry. However, there are other facts which are presupposed in asking the logical question, and to be logical, inquiry must be consistent with all these facts. The problem of the first paper is to determine what *species* of inquiry is rational or logical inquiry. Before Peirce attempts to specify the essential rules of inference, he tries to show the general condition required by the notion of being *subject to rules which are binding upon all.* We might conduct an inquiry by ignoring the way others inquire (Tenacity). We might try to conform to the way some particular limited community makes the passage from doubt to belief (Authority). A more respectable method "from the point of view of reason: (12:10) is to fix our beliefs in accord with the internal principles of human reason itself *(A priori).* The method is to be applicable for an indefinite community. Nevertheless, the failures of this method are the most conspicuous. For while it promises to deliver the mind from the arbitrary determinations of a particular group, it reverts to the "taste" (12:11) of the inquirer. Peirce arranges each of the methods of inquiry as minimal steps of commitment. However, none of the nonscientific methods presupposes enough to meet the logical condition that inquiry must be governed by some rules binding upon all. To meet this condition a further hypothesis is required. There must be some external reality which determines the opinion of any inquirer. The reality hypothesis is a minimal and indispensable requirement if inquiry is to be logical.

For the logical question, the question of validity, to be capable of solution there must be some standard which enables us to distinguish good and bad reasoning. Without the reality hypothesis this is impossible. None of the methods except the scientific enable one to distinguish "a right and wrong way" (12:12). On each of the three non-scientific methods there is no way, by the particular method, to settle disputes among those who use the same method. The standard in each case is not sufficient to make the distinction of correct and incorrect rules, for the great logical distinction of valid and invalid cannot be made without an external standard.

Peirce says in the fourth section of the second paper, "Now, as we have seen in the former paper, the ideas of truth and falsehood, in their full development, appertain exclusively to the scientific method of settling opinion" (12:298). He nowhere explicitly says that only the scientific method presents the distinction of a right and wrong way of inquiry. For if the rules which govern the settlement of opinion on the *a priori* method, for example, are subject only to the inclination of the individual inquirer, two different inquirers may be led from the same premises to contradictory conclusions. If that method furnishes no way to decide whose thinking is right, or what amounts to the same thing, if both are equally right, then there is no way to tell which of the different conclusions is right. In Peirce's view, the claim that only the scientific method provides a distinction of a right and wrong way implies that it is the only method in which the distinction of true and false belief is fully developed. That Peirce saw this implication is made clear when he concludes his discussion of the failure of nonscientific methods to develop fully the ideas of truth and falsehood in the fourth section of the second paper. He concludes that discussion with the statement,

> In contenting themselves with fixing their own opinions by a method which would lead another man to a different result, they betray their feeble hold on the conception of what truth is (12:299).

Only the scientific method with its reality hypothesis is able to make the great logical division of valid and invalid. Only on this hypothesis can we suppose that the passage from doubt to belief is subject to some rules which are binding upon all.

According to this interpretation Peirce is mainly concerned with the problem of validity and especially with that of determining the most general guiding principles —the essential rules— of reasoning. To ask whether any rule is a good rule, that is, to ask whether an argument is such as generally to lead from true premises to true conclusions presupposes that the passage from doubt to belief is subject to some rules which are binding upon an indefinite community of inquirers. Before Peirce attempts to give the essential rules which are binding upon all, he first argues that the reality hypothesis is necessary to guarantee the possibility of valid reasoning. This is the main point of the first paper.

Before Peirce turns to the development of the essential rules of reasoning which will make our ideas true he asks, how are our doubts to be clarified so that they can be resolved by the scientific method? In the first section of the second paper he argues that the need for something better than the clearness and distinctness of the logicians is relative to the demands of the method of inquiry to be used to settle opinion. The main point of the second paper is the development of the rule to attain the grade of clarity which is required for questions to be settled by the method of science. The only other point which might be regarded as the main point of the second paper is the clarification of the idea of reality which is given in the last section. He begins the section by saying, "Let us now approach the subject of logic, and consider a conception which particularly concerns it, that of *reality*" (12:297). The reason that the conception of reality particularly concerns logic is to be found in the first paper. It is an indispensable conception to make the logical distinction of valid and invalid. The general problem of probability is the problem of validity conceived quantitatively, how much are given premises worth to prove a conclusion (12:206)? In the fourth paper Peirce tries to show how synthetic inference can have any worth at all. In both the clarification of the idea of probability and in the explanation of the validity of synthetic reasoning he uses his clarification of the idea of reality.(10) This idea, which is first introduced in the first paper and clarified in the second, forms part of the argument of the whole series in the third and fourth papers. In the third paper he gives all the fundamental rules for calculating the worth of premises which may bear on a given conclusion and in the fourth paper he gives the general rule of valid synthetic reasoning. In the fifth and sixth papers he gives the essential rules for the conduct of induction and of hypothesis respectively. My interpretation of the first paper reveals a definite unity of purpose guiding the composition of all of the first four papers and parts of the fifth and sixth papers into one coherent line of argumentation. It shows that the first paper is the first step of that line by making the reality hypothesis essential to the solution of the logical problem. The essential rules of reasoning which are presupposed in asking the logical question are given in the last four papers.

IV

If I could now show that the main points of the fifth and sixth papers form parts of the argument, I could rest my case. For I would then have shown that each of the six papers is devoted to one line of argumentation which runs through all of them. Unfortunately, I cannot show this to be true of the fifth and sixth papers and the discussion required to show to what extent they contribute to the argument is too lengthy to be given here. What follows is an extremely brief summary of the results of pages 28-46 and 70-88 of the thesis to which I refer in note (1) above.

There are two obstacles to showing that the main points of the fifth paper form part of the argument of the series. The first is that it is not clear that Peirce is mainly concerned with the logic of science in the fifth paper. Two topics are discussed: What bearing does our view of the order of Nature have upon religion? And what bearing does it have upon the logic of science? The discussion of the former, which I call the religious topic, is not part of the argument. While it is a mere aside when the fifth paper is considered in the context of the rest of the papers, this is not clearly the case when it is considered independently of them. I do not have to concede that the religious topic clearly dominates Peirce's attention, but I do have to admit that it is not manifestly of minor importance. Accordingly, I cannot clearly show that the fifth paper is mainly concerned with the argument of the series.

If I could show that all of the logical topic is discussed for the sake of the argument of the series I could claim that the fifth paper is devoted to the argument in so far as the subject of the series, the logic of science, is concerned. Another obstacle is in my way. Much of Peirce's discussion of the logical topic is too unclear to determine what, if any, are the main logical points of the paper. Now there is rather clear evidence that Peirce intended to discuss the logical topic for its importance to the argument of the series. In the last section of the fourth paper he claims that the true theory of synthetic reasoning should be "intimately connected" (12:717) with a general philosophy of the universe. When he announces what I have called the logical topic at the end of the first section of the fifth paper, he says, "What sort of a conception we ought to have of the universe, how to think of the *ensemble* of things, is a fundamental problem in the theory of reasoning" (13:204). It would appear to a careful reader that Peirce is going to develop further the theory of reasoning given in the fourth paper by showing in the fifth the general philosophy of the universe with which the theory is intimately connected. The problem is that much of the discussion of the logical topic is not sufficiently clear to be sure that Peirce carries out this program.

In addition to the specification of the essential rules of induction and the argument against Mill's theory which attempts to explain the validity of induction by the principle of the uniformity of Nature, there are two other points in the fifth paper which clearly form part of the argument. In the second section Peirce completes one of the arguments against the conceptualistic theory of probability. For he there argues that a purely chance world, to which this theory is committed (12:714), is logically impossible. A purely chance world, in which valid synthetic inference would be of no value (12:714), is itself impossible. Any world that may exist has some uniformities and while the uniformities of Nature do not guarantee the validity of induction, they exist in any world in which such inferences may be made. After Peirce discusses the essential rules of induction and argues that its validity is not to be attributed to the uniformities of our world, he tries in the fourth section to show how we may come to know these uniformities once we have made valid statistical inductions. The *problem* with which Peirce deals in the fourth section arises because of the development in the second and third sections of the theory of reasoning which was developed —or begun— in the fourth paper.

While I do not claim that these points which form part of the argument are the "main" points of the fifth paper, I do not have to admit that they are of minor significance. On the contrary, if one supposes that there are main points to the paper, the argument that any logically possible world has some uniformities and the argument against Mill's theory with the specification of the essential rules of induction would have to be considered of major importance. However, in the absence of a thorough analysis of the fifth and perhaps of the fourth papers, which might clarify much of Peirce's discussion, I can only claim that the parts of the argument of the series as a whole which are found in the fifth paper in no way appear to be of minor significance.

There is rather clear evidence that Peirce intended to continue the argument of the series in the fifth paper by exhibiting the general philosophy of the universe with which the theory of reasoning developed in the fourth is intimately connected. This intention was partially fulfilled, but I cannot show that the fifth paper is mainly concerned with fulfilling it because of the discussion of the religious topic and because of the unclearness of the discussion of the logical topic.

There are two main points of the sixth paper. One is the argument of the third section in which Peirce states the essential rules of hypothesis. The other is the elucidation of the distinction between induction and hypothesis which is discussed in the other four sections. The first point is that part of the argument which we have already noticed, and much of the discussion of the second is essential to the argument of the series. In fact, almost all of the sixth paper appears to form part of the argument. The distinguishing of the various kinds of inference in the first two sections is essential to the argument. For the theory of reasoning in the fourth paper clearly presupposes the distinction of deductive and synthetic inference, and the distinction of induction and hypothesis is assumed in the specification of the essential rules of these two kinds of synthetic inference. After Peirce briefly argues that the validity of hypothesis is not based on the uniformity of Nature and states the essential rules of hypothesis in the third section, he illustrates in the fourth the importance of the two kinds of synthetic inference in actual scientific work. Only the fifth, the last section, does not seem to contribute in a significant way to the argument of the series.

In the last section of the paper, and of the series, Peirce attaches great importance to the induction-hypothesis distinction beyond its value to the argument and it is not easy to see how this last section adds much to the argument. Consideration of this section provides a way to construct a plausible case that in the sixth paper Peirce is not as concerned with the argument of the series as I have just suggested. Furthermore, the case might be buttressed by noticing that both the deductive-synthetic and induction-hypothesis distinctions should have been delineated *before* the sixth paper.(11) Because of the importance Peirce attaches to the induction-hypothesis distinction and because the elucidation of the various kinds of inferences is out of place in the sixth paper, I cannot claim that all or most of it is put forth for the sake of the development of the argument of the series. I can only claim that one of the main points of the sixth paper is a crucial part of the argument and that most of the paper in fact deals with issues which are essential to it —though some of these points should have been developed earlier for the most cogent presentation of the argument.

We have seen that all of the first four papers are devoted to the argument which runs from the first to the last papers. The fifth paper considers religious issues which are extraneous to the argument and not all of the logical issues of the paper have been shown to be parts of the argument. In the last paper Peirce attaches such importance to the induction-hypothesis distinction that I cannot say that all of the paper is part of the argument. Furthermore, some of the issues in the sixth paper which are essential to the argument are out of place there. While the parts of the argument found in the last two papers are not merely incidental points when these papers are considered independently of the rest of the papers, it is clear that great unity of treatment of the logic of science in the first four papers is not carried through to the end of the series.

V

We have noticed that the term 'illustrations' in the title suggests a random disconnected treatment of the logic of science. Yet by tracing a line of argumentation through all six papers, and by indicating that the various parts of the argument found in the several papers are not incidental minor points when the papers are considered individually, we have seen the suggestion is not accurate. Why then, we may ask, does Peirce use the term 'illustrations' in his title? Insofar as an explanation can be given from a study of the papers, it is this: according to Peirce, actual work in science illustrates the logic of science. We have seen that in the first sections of each of the first three papers Peirce says or implies that logicians should look to actual scientific work in order to learn lessons in logic.(12) Peirce is quite clearly taking actual works in science as illustrations of the logic of science.(13) I shall discuss only the most important one.

We have already noticed that in the first section of the first paper Peirce concludes his list of great works of science which provide logical lessons by discussing the work of Darwin, Clausius, and Maxwell. Darwin attempted to extend to biology the same method that the latter two had used in the theory of gases. Peirce attempts to extend this method to the logic of science generally.(14) As we have seen, in the second section of the third paper Peirce claims that the problem of probabilities is the problem of logic conceived quantitatively. How much are premises worth as evidence for the conclusion? The great lesson which the work of Darwin, Clausius and Maxwell illustrates is that nothing can be said about the validity of only a particular inference. Peirce stresses that in regard to the individual inference, considered by itself, probability has no meaning and there is no sense in reasoning in the isolated case (12:607, 12:608, 12:609-612). Our only security in making valid synthetic inferences is that we know that these inferences are such as to lead us to true conclusions in the long run (12:718). Just as Darwin could not say what would be true of the individual organism (considered by itself) and Clausius and Maxwell could not say what would happen to the individual molecule, so Peirce cannot say that the individual synthetic inference is valid; he can only say that in the long run valid synthetic inferences will lead us to the truth. The lesson to be learned from the last illustration of the logic of science given in the first paper is the general character of synthetic reasoning.

Whatever other motives Peirce might have had for his use of the term 'illustrations' in the title, study of the text provides an adequate explanation for its use. Peirce believes that logicians have failed to do their job well because they have neglected to learn lessons from the work of scientists. Peirce thinks that he is different. He takes his clues for his account of the logic of science from the great works of science which illustrate the logic of science. While the term 'illustrations' suggests a random treatment of the subject, the explanation of its use is quite in harmony with the actual unity of the series.

There is another point about the title worth noticing. Peirce is illustrating the logic of science. What, in Peirce's view, is distinctive of the logic of science? My reader who has come this far should have no difficulty answering the question. The logic of science is the logic of synthetic inference. A study of the whole of logic would include the study of deductive inference, but the series presents a study of only synthetic inference. It is this part of logic which is the subject of the series. The distinguishing feature of the scientific method of inquiry is that it enables us to make our ideas true, and this is done by valid induction and hypothesis. Deduction may be of great use to a scientific investigator but it adds nothing to real knowledge (12:712, 714, 717, 718). This kind of reasoning is the essence of the logic of science.(15)

When the "First Paper — The Fixation of Belief" and the "Second Paper —How to Make Our Ideas Clear" are studied as they were originally published by Peirce, it will be seen that they had their inception in his attempt to solve what is now called the problem of induction. The present essay suggests that whatever is pragmatic about these first two papers of the "Illustrations of the Logic of Science" will prove to be but a preamble to the validity of synthetic inference. I shall consider only what is now called the pragmatic maxim. By sketching the argument of the series as a whole we have seen that the maxim, the rule for attaining the third grade of clarity, is used to clarify the idea of probability (12:607), and this in turn is used to give an explanation of the validity of synthetic reasoning (12:615, 13:211), and this is used to give the essential rules of induction and hypothesis (13:211, 475).(16) The pragmatic maxim then is a preamble to the validity of synthetic inference in the sense that it is a prelude to Peirce's attempt to solve the problem of induction. It is also a preamble to the validity of synthetic reasoning in quite a different sense. It was designed to be used for the purpose of making valid synthetic inferences. The first two grades of clarity are quite sufficient for the purpose of deduction. However, to add to our real knowledge we must use valid synthetic reasoning, and here is where the first two grades of clarity fail. They do not provide a way for us to learn something new (12:287-288). To do that we must first obtain a "far higher grade" (12:289) of clarity so that we can make our ideas true by valid synthetic inference (12:302).(17) As Peirce's main purpose in the series is to explain the validity of synthetic reasoning, "the great problem of logic" (12:606, see also 12:717), and as the value of the pragmatic maxim lies in its use for making valid synthetic inferences, it appears that we may have to study the maxim in the context of Peirce's theory of synthetic reasoning to obtain an accurate understanding of his conception of the maxim.

The reader who is already familiar with some of the literature on the birth of pragmatism may already see the possibility of correcting some particular commentary or of settling a particular controversy by studying Peirce's original public statement of pragmatism in the context of his theory of synthetic reasoning. Whether my reader sees this or not he should agree that we are not likely to have an accurate and thorough study of the origin of pragmatism until we have such a study of Peirce's attempt in 1877-78 to solve the problem of induction.

NOTES

1. The first four papers were published in volume 12 and the last two in volume 13 of *The Popular Science Monthly.* I shall refer to these papers by giving volume number, colon, page number. Most of my citations will be given in the text without a footnote number.

This essay is essentially a summary of my unpublished M.A. thesis which is on file at the University of Illinois Library (Urbana, Illinois). There my arguments are generally much more developed than they are here, and there I take space to deal with objections which I cannot do here. The only point which is more developed here is the claim that to obtain an accurate and thorough understanding of the origin of pragmatism we should consider Peirce's theory of the validity of synthetic reasoning. This claim and the argument for it are only implicit in the M.A. thesis.

2. The texts were first republished in 1923 by Morris Cohen in *Chance, Love, and Logic* (Harcourt, Brace, and Company). Volumes 2, 5, and 6 of the *Collected Papers of Charles Sanders Peirce* contain edited texts of all six papers. These volumes were edited by Charles Hartshorne and Paul Weiss in 1932, 1934, and 1935 respectively. The order and location of the papers may be ascertained by consulting the editors' note on page 223 of volume 5. In 1957 Vincent Tomas published the edited versions of Hartshorne and Weiss in *Charles S. Peirce; Essays in the Philosophy of Science* (The Liberal Arts Press).

3. See, for example, Arthur O. Lovejoy's essay "What is the Pragmaticist Theory of Meaning? The First Phase" in *Studies in the Philosophy of Charles Sanders Peirce* (Harvard University Press, 1952) edited by Philip P. Wiener and Frederic H. Young.

4. For example, *Introductory Readings in Philosophy* (Charles Scribner's Sons, 1962) edited by Marcus G. Singer and Robert R. Ammerman, contains only the edited text of the first paper.

5. Any reader who doubts this may satisfy himself by reading the second

section of the second paper. The rule for attaining the third grade of clarity is that which is now called the pragmatic maxim. Neither 'pragmatism' nor 'pragmatic' occurs in any of the six papers.

6. My purpose in this essay is not to evaluate Peirce's work. My purpose is to show the program which Peirce attempts to carry out, not to decide whether he successfully carries out his program. For convenience I shall often omit such qualifying phrases as "according to Peirce's view," "Peirce tries," "Peirce thinks," etc.

7. The rule for direct calculation (12:612-3) is not summarized in the fourth paper. The "additional" rule may not be the rule of balancing reasons stated on 12:709; it may be the formula $m+1/m+2$ stated on 12:712.

8. I refer to the line of argumentation of the series as a whole by simply using the term 'argument'.

9. For purposes of this essay it is not important whether or not Peirce does actually use the concept of continuous quantity in his clarification of the idea of probability. I mention his attempt to make use of an idea which he claims to have obtained from actual scientific work because of a point that I want to make later. We shall see that Peirce takes various works of scientists as *illustrations* of the logic of science.

10. Compare 12:607-608 and 12:718 with 12:299-300. I do not intend to say that the way Peirce uses the clarification of the idea of reality is clear in these cases. That he attempts to make use of the clarification is clear.

11. Peirce does give a brief statement of the deductive-synthetic distinction at the beginning of the third section of the fourth paper (12:711); nevertheless, the more detailed discussion of this distinction in the sixth paper would have been more appropriate in the fourth paper —or even earlier— than in the sixth paper. There is no hint of the induction-hypothesis distinction before the sixth paper, although the distinction is assumed in the third section of the fifth paper. The fact that the distinction is not made until the sixth paper forms the basis for the strongest objection that might be raised against the present essay. I deal with this objection on pages 34-45 of the thesis to which I refer in note 1 above. I remind my reader that I do not deal with possible objections in this essay.

12. See pages 41-42 of this essay. Also notice note 9.

13. There are some lessons of logic in the first section of the first paper which I have not mentioned at all, and which Peirce uses in developing his logic of science.

14. I remind my reader that the general problem of probabilities is simply the general problem of logic conceived quantitatively. See page 41 above.

15. For a fuller treatment of this subject see pages 90-96 of the M.A. thesis referred to in note 1 above.

16. See pages 38-40 above.

17. The way in which the pragmatic maxim constitutes a first step in the use of the logic of science is discussed on pages 94-96 of the M.A. thesis referred to in note 1 above.

Southern Methodist University,
Dallas, Texas.

Hartshorne, Merleau-Ponty and Synaesthesia

by George W. Linden

Those of us who were Max Fisch's graduate students in the early fifties were privileged to work our way through(1) Elijah Jordan's major metaphysical work, *The Aesthetic Object.*(2) It was a most difficult experience. But like many difficult experiences, it was richly rewarding. My task in the seminar was to investigate and report on the theory of synaesthesia proposed by Charles Hartshorne.(3) Besides explicating Hartshorne's thesis, I argued two assertions: 1) that understanding Hartshorne's work was a necessary, though not a sufficient condition, for understanding that of Jordan(4) and 2) that the rise of Gestalt psychology would provide evidence for the truth of Hartshorne's basic theory. The first argument may be validated simply by reading Jordan's *Aesthetic Object.* The second I wish to substantiate here.

The theory of a common sense is not a new one in philosophy.(5) It goes back at least to the speculations of Aristotle.(6) In fact, were one willing to assume that Anaximander's obscure sayings are not descriptions of a cosmogony, but a type of introspective psychology, one might interpret his theory of the separating of opposites out of the "boundless" as a description of the dynamics of sensation. To speculative intelligence, it seems clear that what a sensation is is a focal feeling.

If one wishes literary evidence, he merely has to turn to the poets. He will find them using synaesthetic images as the basis of metaphor. Almost every poem of Rilke's later period begins with an analogical definition based on sensations and one could not find a more brilliant fusing of eye and ear than in the lines from Lorca's "The Brawl":

> "Where slithered blood begins to moan
> the dumb song of an injured snake".(7)

While at least one British critic, William Empson, classifies literary synaesthesia as an aberration based on some "obscure physiological perversion"(8) it is difficult to believe that the vast majority of poets of all ages have suffered from the same nameless physiological defect or have taken LSD. The assertion merely proves that Empson should read more poetry.

Continental critics are also incorrect, I believe, when they argue that modern poetry begins with Baudelaire's famous sonnet "Correspondences". This overlooks the fact that Wordsworth, Crashaw, Smart, and Shelley preceded Baudelaire as did many others. Roy Campbell's hypothesis, based unlike the continental critics, more on history than admiration, seems much more acceptable. He states:

> But surely the first man to make these 'correspondences' was the first man who ever spoke of 'sweet' sounds of music, thereby making a contact between the sense of taste and that of hearing —and he must have lived very long ago.(9)

Even Homer's phrase, "the rosy fingered dawn", which has long been a cliche, fuses the lightness of touch with the softness of rose. A survey of poetry will not only discover a rich mine of intersensory analogy, but it will also reveal that much so-called "modern" poetry is indeed ancient.

But let me put aside, or bracket, the history of philosophy and the history of literature with respect to theories of synaesthesia. As fascinating as it may be, such an investigation is not my task here, and in any case, it has already been partially accomplished.(10) My task here is briefly to outline Hartshorne's theory concerning synaesthesia and to introduce more modern evidence for its truth. The modern evidence will come from the interpretations of Gestalt psychology made by Maurice Merleau-Ponty.

Dr. Hartshorne no doubt wrote his book for the same reasons that most philosophers write books: he was sensitive and confused. After all, one mark of sensitivity is the ability to answer obscure questions, and the work of most psychologists in the field of sensation did not agree with Hartshorne's experience. Furthermore, he found their theories to be offensive(11) and irrational. Hartshorne thus proposed his thesis in order to make sense of his own experience and to insert some rationality into the structure of psychological theory. The question arises, then, what were the psychologists doing which Hartshorne thought was irrational? They were slavishly following Helmholtz.

Helmholtz had claimed that sense qualities were irreducible, individual, heterogeneous, and incapable of comparison. He had further asserted that sensory awareness and its contents must be separated and that affection and sensation should be regarded as separate and distinct.(12) Now Hartshorne believed these dichotomies were artificial and unjustified, not only in view of his own experience, but because of modern experiments. Furthermore, he thought they violated the rule for simplicity of structure by emphasizing the differences of sensations and ignoring the basic homogeneity present in all noetic experience. Hartshorne wished to correct error, then, but he also presented his hypothesis with the purpose of unifying the investigations in the fields of sensation and perception and forming them into one rational structure. For these reasons, he proposed a relational view of sensation and perception with a matrix upon which all qualities converge.

Hartshorne's thesis is an attempt to establish five main principles, the first of which is that colors and sounds are not distinct and separated by impassable barriers as had previously been supposed, but that they are closely related and in some cases tend to shade off into each other. He also maintains that sensory qualities shade into affective tones and that they are, in some cases, identical. Furthermore, he argues that there is no difference in kind between sensory awareness and its contents.(13)

His second point is that perceptibles cannot be divided into merely primary and secondary qualities, but that there are tertiary qualities which are as important as the primary and secondary qualities. It is the tertiary qualities which make the others intelligible. He calls these tertiary qualities "aesthetic values" or "affective tonality". This affective tonality is similar to Whitehead's "feeling-value" and Hartshorne claims it is the stuff of consciousness. It is not merely associated with primary and secondary qualities, but is in a sense identical with them. Therefore, Hartshorne states: "Thus the gaiety of the yellow (the peculiar high specific gaiety) is the yellowness of the yellow."(14) Affection does not mean emotion here, as in the James-Lange theory, but denotes feeling, a term which has several dimensions, one being pleasantness-unpleasantness.(15) Hartshorne believes the dichotomies perpetuated by former theories (primarily associationist and S-R psychology) were due to their unsocial view of mind and their failure to apply the concept of continuity to the science of perception.(16)

In the third place, Hartshorne wants to emphasize that experience is basically social in all its aspects or modes. There is a social "otherness" in any perception because everything seen is always seen in relations. He thus states: "The 'coldness' of the green, the 'distance' of blue, the 'aggressiveness' of red embody modes of variation fully explicable only in terms of experience conceived as a social continuum."(17)

The fourth principle is that the intrinsic nature of sensory qualities express organic attitudes. Thus, they tend by their very nature to give impetus to modes of behavior which are appropriate to them in respect to their situations or the situations and relations which they express. Thus the meanings attached to qualities may be explained by the biological adaptiveness related to them. Blue, for example, is appropriate to sky, distance, bodies of water, and short wave-lengths of blue are easily perceived by humans as appropriate to the adaptation of the human organism to the major objects of nature.(18) (Recent biological observations show that some other animals react in similar manner, e.g., robins and stickleback fish become aggressive in the presence of red.)(19)

Hartshorne's fifth, and last principle, is that qualities are evolutionary. He does not view qualities as emergent, but as evolved and related to previous states although they may contain new or different aspects which were not perceptible in the primitive quality which was their origin. He views the present state of a given quality, then, as a development of and from a former more basic quality "rather than as a sheer displacement of the old and irruption of the new".(20)

The five principles of Hartshorne's theory are not to be viewed as five distinct and separate assertions, but rather as forming an organic whole with inter-relations between the various factors. He does not claim infallibility for his theory. But he does hope that it will open up the fields of sensation and perception to new investigations. He also hopes that it will help to explain what had hitherto been deemed "ineffable". To sum up then, Hartshorne proposed a theory "of the contents of sensation as forming an 'affective continuum' of aesthetically meaningful, socially expressive, organically adaptive and evolving experience functions".(21)

Of all the terms used to describe intuition, such as "cognition", "striving", "willing", "knowing", "sensing", there is only one, Hartshorne believes, which throws real light on the nature of the identity and differentiation of qualities and that is "feeling". He not only defines feeling as "the essence of being"(22) but claims that it is the basic function of the intuiting mind by which and through which the other functions may be understood.(23) Now there is a common ambiguity of the word "feeling" as used in our language.

This is due to the fact that feeling embodies three elements: the simple qualitative givenness, striving or abstracting, and the relation of meaning or cognition. Nevertheless, all immediate feeling is feeling "of" some state of affairs or situation which is related to the experiencing subject. Using feeling as his basic matrix, Hartshorne goes on to examine some of the errors which he believes previous theories have committed.

He points out the absurdity that most color theories fall into, for on the one hand, they assume that colors are in some ways homogeneous and in other ways absolutely different. Thus it is claimed that orange resembles yellow, and orange resembles red, but yellow and red do not resemble each other at all (except, perhaps, in the fact that orange is related to both). Yellow and red are held to be absolutely individual and different. The most that could be said for red and yellow on this view is that they share some basic attribute such as "coloricity". One of the obscure questions Hartshorne then raises is this: "how can a continuous graduated difference at one point become an absolute difference at another? "(24) especially when we know that there is a continuous gradation of saturation in each color which we sense as hue.

The reason previous explanations have fallen into this difficulty, Hartshorne believes, is that they have attempted to geometrize the relations of colors and in doing so have overlooked the primordium of feeling and the presence of tertiary qualities. If one takes tertiary qualities into account, then the relation between red and yellow becomes a close one, for what they share in common is warmth. And if it is seen that we can make a rational structure of colors by grading them according to their affective tonalities, then it is possible that other sense modes may be as close to some colors as certain colors are to other colors. The excitement and warmth of certain trumpet notes may be closer to red than yellow and perhaps closer than red is to yellow. Accordingly, Hartshorne constructs his own color chart on the basis of affective tonality and the primary colors become scarlet, buttercup yellow, blue-green and violet.(25) This chart, he claims, is much more rational than the commonly accepted diagrams and is capable of making sense of the affective continuum.

Now if we perceive the redness of yellow through its essential warmth, what does this tell us about the supposedly separate and distinct categories of sensation and affection? It prompts us to believe that they are not separate and distinct categories, but are closely connected. Quoting the experiments of Nafe and others,(26) Hartshorne attempts to demonstrate that there is an essential continuity between affection and sensation. Thus he claims that when we place our hand in a bowl of warm water and constantly increase the temperature, there is an imperceptible transition or passage from the feeling of warmth to the sensation of pain. The pleasant feeling of warmth has now changed into something more intense and focal: pain. Such experience reveals an essential continuity. Because of this inherent continuity in the experiencing, Hartshorne even goes on to argue that the affection is not only in me, but is in the water, that it is an external, socialized feeling, and not merely a comment on what is happening to my hand. Such simple experiments, Hartshorne believes, make the "whole system of discontinuous classes, together with its corollary of the 'ineffability' of sensations, begin to crumble before our eyes".(27)

Next, Hartshorne examines the characteristics which seem appropriate to different sensations to find if qualities can be described as identical. In order to do this, he examines the relations and characteristics of pitch and brightness.

The very word, "light", reveals the nature of colors in their two-fold aspect, for it not only means that a lighter color is less intense, that is, has a different degree of saturation in respect to its fellows, but it is lighter in weight, or less heavy. Odors also appear to present identical phenomena with some being light, spiritual and ethereal, and others, heavy, dense and earthy.(28) Since brightness and elevation are applied to the same things, logic demands that one find their connecting link. Just as "brightness" is applied to sound, so is "loudness" applied to colors.(29) It might be objected, of course, that people just talk in a peculiar manner. But, Hartshorne would reply, why do they talk in this peculiar manner and not in some other? The reason, he feels, that brightness and intensity are applied to both colors and sounds is that brightness is experienced as identical in each case. Hence, following the experiments of Hornbostel, Boring, and others, he concludes that brightness is identical in all sensory qualities.

A theory based on the heterogeneity and separation of sensations from affections cannot properly account for inter-sense analogies and metaphors. Hartshorne examines several of these metaphorical allusions to show that his theory of continuity can explain what would otherwise be regarded as mere literary puzzles.(30) Thus he examines the poetic phrase, "the silver needle notes of a fife"(31) and states that a sharp object is not only thin, but intense, and that the word "silver" captures the brightness and intensity of fife notes.(32) Colors can also be describable in relation to sounds as in "the moon a tinkle of ice in a cool blue glass",(33) and people often speak of hot perfumes or the sultry sweetness of smells. Where there is likeness, Hartshorne argues, there must be more than mere difference.

The explanation for these phenomena may readily be found, by introspection, he believes, to rest upon a common basis. Introspection for Hartshorne is not purely subjective since he wants the term also to include "race observations".(34) When one unites race observations and introspection, he argues, fusion of the sense qualities is found not to be the exception, but the rule and the simple qualities which form a partial identity have a common unifying basis: feeling tone. He describes feeling tone as not extrinsic but intrinsic to the qualities themselves.(35) Thus, the first, is, in a sense, the last, for the tertiary qualities are fused throughout the other categories. This does not mean that qualities have a mere accidental togetherness nor that they cause the perceiver to associate certain affective responses with a given quality. The affectivity, Hartshorne claims, the feeling is *in* the quality. Sensation is, then, seen to arise from a basis of presentation of simple affectivity.

What Hartshorne is trying to do here, of course, is not merely to explain the common basis of synaesthetic metaphors, but also to argue that tertiary qualities are objective. But, one might argue, if tertiary qualities are objective, and if Rimsky-Korsakoff was convinced that F sharp was strawberry red,(36) then such a judgment should be universal. Everyone should be able to perceive that F sharp is actually strawberry red. But, Hartshorne replies, such a demand is too hasty. One should not confuse the objective and the obvious. To claim that an analogy is objective is not equivalent to claiming that it is "obvious or readily detected and described",(37) After all, we must remember that man is a practical animal and what we observe is "chiefly what practical life has made important, and our ability to report in unambiguous terms what we have observed has many a limitation."(38) Nevertheless, if Rimsky-Korsakoff were right in his assignment of strawberry red to F sharp, that is, if such an analogy is actually based upon an objective, affective continuum which unites such a color and such a sound, then this relation is, in principle, open to observation by anyone. Furthermore, it should be constant if taken in the same social context in which it was originally observed.

Hartshorne affirms that it is not merely strange experiences which exhibit synaesthesia, but our most common ones. When, for example, one is eating an apple, it is extremely difficult, if not almost impossible to try to separate the taste of the apple from its scent, and this taste and scent are also related to the pithy texture of the apple and its concomitant color. When one eats ice cream he does not separate the sense of taste, cold, smoothness, and scent, but experiences them as a whole.(39) In fact, Hartshorne argues, it is one of ice cream's affective constituents, the pain of the cold which, by its subordination and relations to other senses such as taste, makes the ice cream pleasant. No one asks for warm ice cream. The heterogeneity of sense experiences is shot through with similarity and in many cases it is the coexistence and relations of opposites of affectivity which determine the affective tonality of the whole. What is contradiction in logic is contrast in sensation.

Tertiary qualities are aesthetic characters, and just as primary characters have been held to be nothing without secondary qualities, so secondary qualities are nothing without tertiary qualities. The tertiary qualities form a matrix and are the common primordium, Hartshorne argues, to which the secondary qualities, such as color or tone, refer.(40) Thus he states: "the feeling-tone of a color is not, according to an idealism that understands itself, something over and above the color; it is just the color itself seen in its intelligible essence."(41) Furthermore, he does not believe that this rests on mere whim, but that "objective feelings are observed facts."(42) Granted that there needs to be improvement in our methods of observation and that we should be able to construct criteria so that we may be able to separate what really appears in experience and what merely seems to appear,(43) it is to immediate experience that we must return. As Hartshorne puts it:

> The fact is that individuality is identifiable only by direct grasp of a thing as a 'this', that is to say through its relations to an experiential context. The individual and the real is what acts upon and is acted upon by us.(44)

We shall find Merleau-Ponty also urging us to return to immediate perception and a pre-perceptive ground.

According to Hartshorne's theory, then, sensations are evolved affectivities. Nor is it absurd to reason that since certain affectivities have evolved into sensations, that others may do so. "Almost the only thing overlooked", says Hartshorne, "by the distinguished psychologist", (Titchener, whom he quotes on this point), "is that the character separating 'sense' from 'feeling' is not chiefly qualitative difference".(45) The decisive factor is "the manner in which sense quality is localized in the body or externalized, projected out of it",(46) The difference, then, is not one of kind, but of location and context.

The externality of characters, Hartshorne argues, is "a stubborn and irreducible fact, so far as observation goes".(47) He doubts that anyone can deny, in view of experimental facts and his theory that all of the four following relations maintain: "Sensations cannot exist without feelings, sensations can vary without feeling varying, feeling can vary without sensation varying, and that feelings can occur without sensations",(48) Thus he states: "There is but one problem, the qualitative dimensions of a pervasively affective experience with a subjective and objective aspect".(49) And since sensation is now viewed as an evolutionary process emerging from affective unity, Hartshorne emphatically rejects the positions of associational and faculty psychologies. Associationist psychology, he asserts, realizes that some kind of a relation exists between the things associated. But it is completely incapable of accurately describing this relation, how it exists, why it exists, or what it unites.(50) He summarizes Pikler's experiments and his theory of dimensional qualities as evidence for the objectivity of affective tonality. Hartshorne admits that Pikler's theory may be wrong, but he thinks it is the best working hypothesis we have.(51) At any rate, he believes it will help us to resolve some puzzles.

One puzzle is black. Black has all the characteristics of a color. It exhibits hue, gradation, saturation, texture and all of the other characteristics of red or yellow. Hence we commonly say that black is a color. Suppose, however, that we ask, as Hartshorne does, why we see black or at least a dull gray when there is a complete absence of light? Isn't it just as plausible to see blue or green, or even yellow or red in the absence of light? Most orthodox psychologists would say the question is nonsense. They would simply say "black is the absence of light". But this is to define black as a non-color and furthermore to claim it is ineffable. Hartshorne answers that black is indeed a true color and that the answer to why we see black when there is no light is simple and self-evident. "The qualities of these colors are not appropriate to the disappointment of the desire to see. Black is thus appropriate".(52) The intentional nature of consciousness determines the absence of light as the color, black.

If darkness is a puzzle from the perspective of orthodox psychology, so also is light. As Merleau-Ponty has observed, when we first switch on the light it "seems yellow at first but a moment later tends to lose all definite color".(53) Hartshorne would explain the appearance of yellow as the function of joy and the satisfaction of the desire to see. Merleau-Ponty would disagree that the apparent yellow is a function of the operation of intelligence. He argues instead that it is a function of the very "configuration of the field" and that I do not think the world into perception, but find it organizing itself "in front of me".(54) Both would agree, however, that what we see when we turn on the lights is yellow, that this sensation is not an illusion, and that it is not subjectively arbitrary, i.e., capable of being willed away by the subject.

Before his early death, Maurice Merleau-Ponty had already gained the reputation of being one of the major thinkers of our era. He was, in my judgment, the most articulate and important of the existentialists. His two major works are only beginning to exert their influences in this country, but once absorbed, they will have profound effects, especially on psychology. Both of the major works(55) are attempts to review and reconstruct psychology as a gestalt science from an existential point of view. The remarkable thing for our purpose here is that though he started from an entirely different philosophic point of view and different psychological theories, Merleau-Ponty ends up describing a theory of synaesthesia remarkably like that of Charles Hartshorne.

In order to examine the relations of Merleau-Ponty's discussion of synaesthesia in relation to that of Hartshorne, I shall first explain why Merleau-Ponty rejects classical psychology and what he believes are the leading characteristics of the new psychology. I shall then state his account of perception in general and synaesthesia in particular. Next I shall compare his views with the five theses of Hartshorn's hypothesis. Lastly, I shall examine some of the major differences in their philosophic positions.

Classical psychology (and that of Marshall McLuhan)(56) claims that our sensory experience consists of a sum or mosaic of sensible elements.(57) These elements are called qualities and are taken as the given "psychic facts" of experience. While classical psychology is aware that there are relations between sense qualities, it maintains that their unity is a human construction which is due to the operations of intelligence and memory.(58) Thought, therefore, becomes basic and it is by judgment that one holds the neutral data of sensation in present experience until the mind comes to rest on some hypothesis to interpret their significance.(59) The perspective which one takes in perceptive experience is a function of intellectual choice. Furthermore, since there is a split between the mind and the body, one can only perceive the external signs of others and then resort to his introspection to interpret the "psychic facts", e.g., anger, desire, etc.(60) One comes to an understanding of others, then, by perceiving external signs, interpreting them according to one's own introspective experience, and then projecting these interpreted symbols to the other. Merleau-Ponty rejects all of the above assertions and claims that psychology can advance only when it learns to abandon the distinction of the mind and the body and returns to an examination of perceptive and pre-perceptive phenomena.(61)

Perhaps an analogy will be helpful at this point. What Merleau-Ponty calls "classical psychology" can be divided into two types: intellectualist (idealistic) psychology and empirical psychology. Intellectualist psychology is convinced of the existence of a unified transcendental ego which constructs the world as its object. Intellectualist psychology cannot see the trees for the forest. Empirical psychology is convinced of a fluid stream of separate sensations which bear a one to one correlation between the stimulus and the response. Empirical psychology cannot see the forest for the trees. Existential psychology denies that there is a given transcendental ego which constructs the world from the intellect. But it also denies that the self is a passive place through which there occur streams of stimuli and responses. Existential psychology claims that the trees emerge from the vague background of the forest because man finds himself thrown into a world which he lives(62) prior to his thinking about it. Existential psychology claims that the figure of the tree "owes its characteristic appearance to the ground from which it emerges".(63) Existential psychology cannot see the intervals for the trees.

According to Merleau-Ponty, the basic error of classical psychology, whether rationalistic or empirical, is that it came to rest in the constancy hypothesis as dogma. The constancy hypothesis, whether in psychology or physiology assumes there is a given, objective world. Furthermore, it assumes that the object is some type of transmitter which sends signals along special nerve paths to a receiver who then projects the meaning of these deciphered sense data back into the original text of the world. But, Merleau-Ponty objects, even the physiologists who employ the constancy principle realize its purely theoretical character.(64) Even though science tells me that the cones of my eyes have limited areas and chromatic thresholds, what I perceive is an area before me completely colored, not one with gaps. In the case of color, "the 'sensible' cannot be defined as the immediate effect of an external stimulus".(65) What my retina recognizes *is a field* and hence it is groups, not elements, which are primary in perception.(66)

What this means is that perception of the whole is more natural than perception of the elements. Elements are not primary, but are the results of intellectual analysis. What is primary is the figure/ground relationship and because it is primary, one finds himself with certain privileged perceptions. The reason I cannot see the intervals for the trees is that my body, in its immediate coexistence with the world, constitutes the trees as foreground and the intervals between them as background. What is true of objects, such as trees, is true also of qualities: "the apprehension of quality, just as that of size, is bound up with a whole perceptual context".(67) Perceiving the world as composed of discrete elements which are then later synthesized by the intellect under the rubrics of "attention", "memory", etc. is a "belated and rare attitude".(68) The normal functioning of perceptual experience is an active process of integration in which "the text of the external world is not so much copies, as composed".(69) Furthermore, meaning is already invested in the structures of perception and the element is no longer seen as that which occasions the construction of the whole nor as that which by cumulative addition constitutes it.(70) Sensations are, in a sense, artificial products, but they are not merely mechanical nor arbitrary. I perceive the trees *with* my eyes and this "with" is not merely instrumental and we know "that the sensory apparatus is not a conductor, that even on the periphery the physiological impression is involved in relations formerly considered central".(71) It is for this reason that I may be quite familiar with a face and yet unable to tell you the color of the eyes.

The assertion of the constancy hypothesis and the assumption of the elementary as primary in perception are both results of the fact that psychology has swallowed whole the scientific reconstruction of the world. But the scientific world is an intellectual world and man does not think the world, he lives it. Hence, Merleau-Ponty argues, we must realize that our body is a way of being-at-the-world-from-within-it. We must return to the primacy of perception. And when we do so, we shall realize that we have a commerce with the world and a presence to it that is older and more primitive than intelligence.(72) Man is an incarnate being and his perceptual habits are not formed by some pure or passive spirit, but through the interaction of an embodied person with a coexistent world.

What holds for man, holds, analogously for other animals. The stimulus-response arc which has been held as fundamental by most classical psychology is not normal behavior, but on the contrary, close to being pathological: "the behavior of a sick organism".(73) The understanding of behavior and perception is not, then, merely the establishing of certain empirical coincidences, nor is it to establish certain mechanical correlations through mathematics or otherwise. Instead it "is to unite the ensemble of known facts by means of their significations, to discover in all of them a characteristic rhythm".(74) Merleau-Ponty thus concludes: "it is necessary in this sense to go beyond mechanism".(75) But one must be careful here. There is nothing in behavior which announces the presence of a pure consciousness which is independent and rests in itself *(en soi)*.(76) Nor is the behavior of the animal directed toward pure being. It is directed toward "being-for-the-animal",(77) a certain milieu which is characteristic of the species. Thus, while a chimpanzee will stand upright for certain scientists and certain experiments, his tendency is to return to his animal posture and to "manifest a sort of adherence to the here and now, a short and heavy manner of existing".(78)

Just as there is no pure *pour-soi* and *en-soi* at the level of the animal, so there is no such clarity in human experience. Man is an organism with a more complex structure of behavior, but still an organism. And "every organism" said Uexkull, "is a melody which sings itself".(79) Now it is characteristic of musical texture that the elements form a whole and are internally, not externally related in the constitution of the whole. The relation is not one of mere juxtaposition, but one which is "internal and necessary in the wholes".(80)

Since the melody, the musical text and the unfolding of the gestures all participate in the same structure and point to a common nucleus of signification, Merleau-Ponty claims that "the expressive value of each of the three ensembles with respect to the two others is not an effect of their frequent association: it is the reason for it".(81) Man is an expressive animal. And just as notes express themselves through the whole of the melody, contributing to the whole and being enhanced by it, so man expresses himself with his body in relation to his world.

The classical assumption, then, that what one perceives are external signs of internal emotions must also be rejected. The method of introspection, says Merleau-Ponty, "tells me almost nothing".(82) Anger, fear, love, etc. are not "psychic facts" which I discover by an effort of introspection to be residing within me,but are styles of conduct, modes of behavior which are revealed whenever I abandon the effort to coincide with my feelings and regard them as modes of behavior.(83) Likewise, these are not "psychic facts" hidden in the depths of the other, but are expressions of behavior which may be viewed from the outside. "They exist *on* this face or *in* these gestures, not hidden behind them."(84) Meaning adheres to behavior. Small children can understand gestures and facial expressions long before they can succeed in imitating them.(85) Science has attempted to give a mechanistic interpretation of behavior and sensation but this is because it approaches the organism through a set of "snap shots"(86) of the functional body, that is, a set of abstractions. The illusion that my body is a screen between myself and the world is merely an illusion. My eyes are not "an ensemble of transparent tissues or opaque tissues and organs" but are "instruments of my looking".(87) Classical psychology, on the other hand, acts as though man were in an indifferent world which did not touch him.

Since some of Merleau-Ponty's main themes are now emerging, perhaps I should state them explicitly. What he is claiming is that man is an incarnate being. While things are in space/time, man inhabits space-time and he is able to inhabit the world because objects are *for* consciousness. This "for" is possible because man is an expressive being who imports meaning into things as well as discovering it there. Like Husserl, he urges us to return "to the things themselves" and when we do so, we will find that perception is not some type of low-grade thought (as it was for Descartes), but that it is our primary way of being-with-the-world. Things are *next to* things. Men are *with* things. Intentionality is a direct, sensible, emotional, affective access to the *Lebenswelt* or human world. To return to the things themselves is to return to that "world which precedes knowledge of which knowledge always *speaks*".(88) And since we are already engaged in the world, it is necessary to take the stance of ideality in order to know one's experience, for understanding consists of retrospective grasp. Man is a concrete intersubjectivity who has access to the world through his body. But man also has the power to transcend any concrete situation and consequently a study of his incarnate condition is a study not merely of the embodiment of a personal awareness in a body but also social consciousness embodied in institutions. Man brings his past with him into a present which he transforms with a view to a future. The world to be described, then, is not a noumenal world of things in themselves nor a transcendental ego in itself (contra Sartre) but a phenomenal world. A world in which both the self and its objects are never completely given but are always subject to some opacity.

The fact that the human order is based on lived experience, i.e., perception, means that perception is the dynamic ground of meaning. For a child, its mother's face is not a collection of qualities, but a center of expression.(89) The child's perceptions are directed toward human intentions, not pure qualities; the child grasps meaning prior to qualities.(90) This is possible for the child, not because he has some *a priori* set of forms which signify or represent "mother angry" or "mother loving" but because representation is only one mode whereby consciousness can envision its objects. In fact, consciousness is "a network of signifying intentions, sometimes clear for themselves and at others lived rather than known".(91)

While consciousness imports meaning into phenomenal objects, it refuses to be identified with them. Hence it constantly returns to its lived reality. In doing so, it negates its present in favor of the future. What perception reveals, then, are objects capable of supporting a future and man's condition is a condition of basic ambiguity. The self is never completely known and the world is never completely known. Furthermore, in relation to society, man is the result of social structures which he helps bring about but which at the same time imprison him(92) so that his meaning consists in rejecting those conditions and surpassing them. Consciousness is in the habit of ignoring itself and it is this which has caused classical psychology to misconceive consciousness as though it were hallucinated by objects. As Plato correctly pointed out, the one thing common to the gaining of vision and the loss of it is blindness.(93) Consciousness, then, says Merleau-Ponty is intentional consciousness, situated consciousness, consciousness in this body and in this particular world.

If one abandons the classical point of view, which is to say a purely "objective" psychology, then the self ceases to be a mere "where," a bare locus in which sensation happens. That such an abandonment seems necessary is due to the reflection that the world is not perfectly explicit to the perceiver and, in fact, is only displayed gradually. The relation between the perceiver and the perceived, then, is not an abstract one such as a causal relation, but an intimate, constitutive one. The world-for-me is something which is re-created, re-constituted "at every moment".(94) There is an intimate, lived relation between the perceiver, his body and his world. Since this relation is prepersonal and is an interactive relation, the individual's empirical self and body never quite become complete objects just as the world never becomes a completed object. One can never say "I" absolutely.(95) Thus Merleau-Ponty argues that sensation is not a state of consciousness nor a quality of consciousness nor the consciousness of either a state of a quality, but is instead a dynamic means of patterning behavior.

The ground of sensations, it must be remembered, is the body. Furthermore, sensory excitation minus significance has no effect on the body's mobility.(96) Hence a study of sensations should be not a description of simple quale, but of the motor physiognomy, the living significance which colors or sounds institute as they reverberate through the nervous texture of the body. The outline of the self is the inline of the world.

It has long been accepted, even in classical psychology, that the perceptual side and the motor side of behavior are in communication with one another and that this relation is not an abstract or causal one. The problem is to describe this relation. Red and yellow favor withdrawal to the center; blue and green favor going out to the stimulus.(97) "Blue occasioned conduct"(98) is *not* the effect of the wave-lengths impinging on an objective body. Colors are not closed states but are forces which impinge upon my general setting and way of coming to terms with my world. They provide an "atmostphere" for the living self just as music is the patterning of tension felt in the body. The body adopts the quasi-presence of colors, it anticipates them before they become objective spectacles. One anticipates red and then perceives it. Red is not the cause of motor reactions, on the contrary, by its very texture, it is already an amplification of my motor being.(99) To my nervous system, red *is* violence, that's why it signifies it. One's body incorporates red, the self expects red and then finds this expectation confirmed. The subject of sensation, then, is a power born into an existential situation, and if we could learn to relive colors as our bodies do, we would realize that the sensible becomes a certain way of being-in-the-world. The relation between the self and sensible qualities is literally one of communion.(100)

Merleau-Ponty gives us a formula in an analogue to try to grasp this relation by stating that the sentient: sensible:: sleep: slumber. When one goes to sleep, he adopts a voluntary attitude toward slumber. In doing so, the respiration slows and deepens. This voluntary rhythm of respiration continues until it slips into my being. Sleep comes when a voluntary attitude receives confirmation *from outside.*(101) What was significance becomes situation. In the same way, Merleau-Ponty argues, since the relation of intimate co-existence dispels the distinction of the in itself and for itself, qualities cast a spell on the self. The self enters into sympathetic relations with them and they have sacramental value. Sensation is not an invasion of the sentient by the sensible nor is it the case that one acts while the other suffers. What happens is that red and I call each other forth. The red invites me, calls me forth. I must find an attitude which will allow it to reveal itself and then the red gives me back what I have lent it.(102). This, of course, was only what I took from it in the first place.

The process of perception, then, is the process of expressing a certain situation. One cannot say "I see red" in the same sense as one says "I understand the book" or "I have decided to become a professor".(103) The reason is because personal acts create a situation whereas sensation expresses a given situation. Granted that my body has a primordial contract with the world (its peculiar perspective), it still does not follow that sensations are personal. One reason they are not personal has already been mentioned. They do not involve conscious decision and thus sensations run through me without my being the cause of them. But a more important reason is that sensations are always incomplete *and so is the self which perceives them.*(104) The person who sees is never exactly myself for my eyes are merely one specialized portion of myself which happen to be familiar with one sector of being. Since each sensation belongs to a certain field, sensation is restricted. Vision is always surrounded by a horizon of the non-visible.(105) My eyes allow me to be open to the world under a certain aspect, the visible, but they do not give me the whole even though they refer to it. What experience brings is an open totality, not a closed one for I am "thrown into" the world, wrapped up, engaged in it.(106) Only God could perceive the world as a closed totality.

If sensation is not some inert substance, but is a partial process, a way of access to an inexhaustible synthesis, then the unity of the self and the sensible cannot be a real one, but merely a presumptive one. It is the incompleteness of the sensible which accounts for the impersonality and anonymous character of sensation.(107) Since each organ provides one way of access to the world, each provides a particular structure or manner of being in space. Each sense constitutes a small world which opens on the whole. Thus, says Merleau-Ponty, when I open my eyes in the concert hall, the visible space appears cramped.(108) The music besieges visible space(109) for both auditory and visual space lay claim to the whole common world. They are united in their clash.(110) Differences of quality, then, are differences of structure. Vision does not merely add quantity to experience, but alters the structure of the whole of life. To alter the structure is to alter the significance.(111)

Touch, in relation to space, is restricted by the thickness of time. One must touch this before he touches that. Vision, however, perceives two things as simultaneous. Hence touching is not seeing, for touch is not spatial in the same way as is vision. In face, there "is a space so strictly tactile that its articulations do not and never will stand in a relationship of synonymity with those of visual space".(112) Granted that touch is not spatial *as is* vision and that the space of touch is, in some cases, not interchangeable with that of vision, still they must be analogous. Gesture, such as reaching for an object, presupposes such analogy.(113) Touch, then, must be so organized as to make its coexistence with visible space possible.(114) Were this not the case, touch could not communicate with vision but the communication of vision and touch is a fact in the normal adult.(115) It is through the interplay of sensory realms, Merleau-Ponty argues, that space itself becomes observable.(116) Such interplay is possible due to the difference between sensation and natural perception.

Sensation is the result of local vision. When I question my gaze and narrow my field, a sensible quality emerges as an object. A sensory quality, then, is a "peculiar production"(117) of second order vision, the attempt to tell myself what it is I see. Natural perception, however, is achieved by my whole body reacting all at once(118) to an all-at-once world. Natural perception is a world of inter-acting senses, which one discovers when he yields to his gaze and surrenders himself "to the spectacle".(119) Only if one is intellectual and impartial, that is, *indifferent* to his experience, does the constancy hypothesis seem tenable. The closer one approaches natural perception, the more the barriers between the senses collapse. Just as each sense provides its own structure of natural unity, so also, claims Merleau-Ponty, there is a "primary layer of sense experience which precedes its division among the separate senses".(120)

Because each sense reveals a structure of being which can't be exactly transposed, intelligence can only supply an apparent communication between different modalities, nor can an appeal to physiology help. Cell transference will not account for synaesthesia.(121) Yet, claims Merleau-Ponty, synaesthetic perception is not the exception, *it is the rule.*(122) The reason we fail to realize this, he says, is because science has altered our center of gravity in experience. Science has taught us to unlearn "how to see, hear, and generally speaking, feel".(123) We are taught to construct our body image as the physicist and biologist conceive it and then to deduce from this scientific construct what we ought to see, hear or feel. Most men, then, do not live their lives, they *deduce* them. But Merleau-Ponty asks us to put science aside and to return to natural perception where we will find that the senses "intercommunicate by opening onto the structure of the thing",(124) where they each modulate the thing and "communicate through their significant core".(125) Perception, then, does bring us to a single world, but the problem remains of how this synthesis is achieved.

Because Merleau-Ponty has rejected the formalism of consciousness advanced by classical psychology and made the body the subject of perception, we would expect him to provide a bodily base for the synthesis of qualities. This he does, again by analogy. Perception brings out sensations together into a single world in the way that "binocular vision grasps one sole object".(126) The analogy, then, is this: sensory qualities: synaesthetic object: monocular vision: binocular vision. Monocular images appear in front of the object and they slip into unity due to the intention of the subject. But, and here is the important point, though the unity of the object is intentional, it is not "the epistemological subject" who creates the synthesis, "but the body".(127) The body, Merleau-Ponty asserts, is a "synergic system"(128) and it is the phenomenal body, "the prelogical unity of the body image"(129) which provides the unity of the object. Intellect tells me that what I am seeing is the same object; my eyes tell me that I am progressing toward the presence "of the object itself".(130) Once binocular vision is achieved, I view the monocular images as "appearances" of the thing. Even when binocular vision is accomplished, however, I still have not reached the "ipseity"(131) of the thing, for there are still further aspects beyond yet to be perceived. Though unity is achieved, it is never a complete unity. Again, neither the object nor the body image which synthesizes it is given; both are partly open and incomplete. My eyes do not merely coexist, they *collaborate* and hence the unity of experience is "not a formal unity, but a primary organization".(132) My body "is the fabric into which all objects are woven"(133) and the significance, the expressive value of qualities is "the ground of the antepredicative unity of the perceived world".(134) Things "grip"(135) my body before I have symbols or concepts by which to intellectualize them and their qualities are pregnant with reference to each other. Hence Merleau-Ponty concludes that "the senses translate each other without any need of an interpreter, and are mutually comprehensible without the intervention of any idea".(136)

Now that we have a general sketch of Merleau-Ponty's account of synaesthesia, let us return to Hartshorne's theory. His theory included five main theses. We shall review and compare them one by one.

The first thesis was that there are no impassable barriers between sensory modalities. Merleau-Ponty agrees with Hartshorne on this and believes that clinical synaesthesia, i.e., the experiences with mescalin, LSD, etc. establish the truth of this point. Hartshorne had also held that there was no difference in kind between sensory awareness and its contents. Merleau-Ponty agrees with this point also but he insists that the unity is merely presumptive and never final. Furthermore, he distinguishes levels of organization between the various sensible contexts thus explaining their differences on the basis of structure. The unity of sensibles for him, then, is not an identity as it is for

Hartshorne, but an equivalence, that is, the object of vision is equi-valent, has an analogically equal value but not an equal existence to the object of touch.

Hartshorne's second thesis is that affective tonality, feeling-value, is objective and he calls these tonalities tertiary qualities. Merleau-Ponty concurs that meaning is *in* sensory experience and not deduced from it. However, he vigorously denies that such significance is therefore "psychic". In fact, he expressly denies that sensory qualities are reducible to feeling and points out that colors may affect my behavior without my being conscious of them.(137) Sounds, touches, colors, etc. are not closed states but are forces which impinge within me upon my general perspective and way of relating to the world, but the "me" involved is an "impersonal" me. It is my body image.

The third thesis is that mind must be viewed as social and qualities as relational. Merleau-Ponty differs from the other existentialists such as Sartre and agrees with Hartshorne. He views the self as an active synergic, living system and reflection as a precipitate of this living activity. One is constantly surprised by himself as qualities energize the potential of the self. Not only does he argue that mind is dependent on its institutional context, e.g., it is the presuppositions of science which lead us to deduce our experiences, but Merleau-Ponty also holds that the relation between self-world is an intimate one of communion. The same is true for others. In so far as I am with others, I am permeable to their being and they to mine and it is only from reflecting upon the gaze of the other that I can discover who I am.

Hartshorne's third point is that sensory qualities express organic attitudes. Merleau-Ponty agrees and even claims that at the level of symbolization, for example, that of language, that a "word" is then indistinguishable from the attitude it induces"(138) and that hence words have a "physiognomy" to them much like that of persons for we adopt toward them "a certain form of behaviour".(139) My body, he holds, is not merely an object among objects, but that peculiar unique object which "is *sensitive* to all the rest, which reverberates to all sounds, vibrates to all colors, and provides words with their primordial significance through the way in which it receives them."(140) The body uses its own parts as a "general system of symbols for the world" and because it does so, one can "be at home" in the world, "understand" it and "find significance in it".(141)

The last thesis which Hartshorne advances as part of his general theory is that sensory qualities form a continuum and are evolutionary in character. Merleau-Ponty agrees that they are evolutionary, but he denies that there is a continuum. He argues instead that there are emergent levels of function. That which is given to vision is not exactly the same as that which is given to touch. Nevertheless touch and vision intercommunicate for they lay claim to a common world. Not only do they claim a common world, but they provide the possibility for the use of the other, i.e., vision prepares the horizons for the use of gesture.

Merleau-Ponty's development of existential psychology, then, appears to substantiate Hartshorne's theory of synaesthesia. Nor is he the only one to provide such evidence.(142) The main difference between the two theories hinges upon the nature of the synthesis. For Hartshorne, the synthesis is achieved through feeling by mind; for Merleau-Ponty, through the body-image. But perhaps in closing we should note some further differences.

Certainly there is a difference of method. Hartshorne's method is one of intense introspection. Merleau-Ponty denies that one can learn much from introspection and his method is one of phenomenological description. Perhaps more importantly, at least from Hartshorne's point of view, they differ utterly in goal. Merleau-Ponty is out to describe what it means to be human and to find oneself thrown into a world. While he is trying to steer a middle course between the transcendentalism of Kant and the empiricism of British philosophy, he is not primarily trying to argue a metaphysical position. Hartshorne is. Hartshorne's conclusion is that the world in all of its aspects is pansychic and even goes on to argue that it is not an absurdity to assume sentience for single cells and electrons.(143) Although he begins with an account of synaesthesia, Hartshorne's ultimate purpose is theological. He wants to prove that the universe is some kind of mind-stuff. There is no doubt that Merleau-Ponty would reject such theories. To accept them would be to deny the careful analysis of levels of behavior which is the heart of his description of man's condition. Merleau-Ponty would no doubt agree with Elijah Jordan who said: "a pan-psychism or pantheism of process is an interesting spectacle; but good, aesthetically, when you don't need or want to be convinced".(144)

Merleau-Ponty's analysis of sense experience is remarkably fruitful but it also leaves some unanswered questions. One question is the exact relationship between the different sensory modalities. He rightly emphasizes their differences, as Hartshorne seldom does, and also their unities. But beyond telling us that the relation is one of analogy, an analogy of equi-valence, he leaves us in the dark. The other question is why we experience anything at all, why a color, for example, calls me forth from my gaze in the first place so that I focus on it as foreground in contrast to the field. He does give a partial answer in the form of intentionality and hints that the reason a sensation can so occupy consciousness is because there is a common rhythm between the self and the world. But he does not go on to analyze rhythm. To do so would take him into a realm he would rather avoid: metaphysics.

In the end, therefore, we are brought back to the beginning: Jordan's *Aesthetic Object.* Jordan was a metaphysician and the purpose of his work was to analyze the principle of analogical identity and rhythm as the basis of value. Perhaps one need not answer such "obscure questions". But if one does, at least we know where the answers may be found.

NOTES

1. On reflection, I suppose I should say we stumbled through it.
2. Jordan, E., *The Aesthetic Object*, Bloomington, Indiana, Principia Press, 1937.
3. Hartshorne, C., *The Philosophy and Psychology of Sensation*, Chicago, University of Chicago Press, 1934.
4. Hartshorne is the only philosopher he singles out for praise. It would seem that some theory of synaesthesia would be presupposed in order to understand his assertion that: "It is highly doubtful whether one who does not hear a good picture can see it at all as a value object". Jordan, op. cit., p. 117.

5. Some of the history is reviewed in Gleen O'Malley's fine article. See O'Malley, G., "Literary Synaesthesia", *Journal of Aesthetics and Art Criticism*, Vol. XV No. 4 (June, 1957), pp. 391-411.

6. See Aristotle, *De Anima* II, 420 a and b.

7. Lorca, translated in Roy Campbell's *Lorca*, New Haven, Yale University Press, 1952, p. 26.

8. Empson, W., *Seven Types of Ambiguity*, Norfolk, Conn., 1947, pp. 12-14.

9. Campbell, R., op. cit., p. 25

10. See O'Malley, op. cit.

11. I suppose what he meant was intellectual distress, not emotional distaste.

12. Hartshorne, op. cit., p. 6

13. Ibid., pp. 6-7

14. Ibid., p. 7.

15. Ibid.

16. Ibid., pp. 7-8

17. Ibid., p. 8

18. Ibid.

19. Ardrey, R., *The Territorial Imperative*, New York, Atheneum, 1966, pp. 95-6.

20. Hartshorne, op. cit., p. 8

21. Ibid., p. 9.

22. Ibid., p. 13.

23. Ibid., pp. 38-9

24. Ibid., p. 41.

25. Ibid., p. 221.

26. Ibid., pp. 49-56.

27. Ibid., p. 57.

28. Ibid., pp. 59-60.

29. Ibid., p. 61.

30. This seems to be the view taken by Empson.

31. Hartshorne, loc. cit., p. 74.

32. Hartshorne seems to overlook the penetrating subconscious pun involved between "silver" and "sliver".

33. Hartshorne, loc. cit., p. 75.

34. Ibid., p. 74.

35. Ibid., p. 76.

36. Birren, F., *Color Psychology and Color Therapy*, New York, McGraw-Hill, 1950, p. 153.

37. Hartshorne, loc. cit., p. 78.

38. Ibid.

39. Ibid., p. 80.

40. Ibid., p. 97.

41. Ibid., p. 98.

42. Ibid.

43. Ibid., p. 106.

44. Ibid., p. 100.

45. Ibid., p. 115.

46. Ibid.

47. Ibid., p. 118.

48. Ibid., p. 122.

49. Ibid., p. 130.

50. Ibid., pp. 136-7.

51. Ibid., pp. 146-50.

52. Ibid., p. 213.

53. Merleau-Ponty, M., *Sense and Non-Sense*, Northwestern University Press, 1964, p. 51.

54. Ibid.

55. The two major works are *The Structure of Behavior* and *The Phenomenology of Perception.*

56. McLuhan, M., *Understanding Media*, New York, McGraw-Hill, 1964.
57. Merleau-Ponty, loc. cit. p. 48.
58. Ibid., p. 50.
59. Ibid.
60. Ibid., p. 52.
61. Ibid., p. 53.
62. Ibid.
63. Merleau-Ponty, M., *The Structure of Behavior*, Beacon Press, 1963, p. 23.
64. Here he refers to the psychologists Stumpf and Koehler. Merleau-Ponty, M., *The Phenomenology of Perception*, New York, Humanities Press, 1962, p. 7.
65. Ibid., p. 8.
66. *Sense and Non-Sense*, p. 48.
67. *The Phenomenology of Perception*, p. 8.
68. *Sense and Non-Sense*, p. 49.
69. *Phenomenology of Perception*, p. 9.
70. Ibid., pp. 9-10.
71. Ibid., p. 10.
72. *Sense and Non-Sense*, p. 52.
73. *The Structure of Behavior*, pp. 43-44.
74. Ibid., p. 158.
75. Ibid.
76. Ibid, p. 159.
77. Ibid., p. 126.
78. Ibid.
79. Ibid., p. 159.
80. Ibid., p. 121.
81. Ibid.
82. *Sense and Non-Sense*, p. 52.
83. Ibid.
84. Ibid., pp. 52-3
85. Ibid., p. 52.
86. *The Structure of Behavior*, p. 205.
87. Ibid., p. 219.
88. *Phenomenology of Perception*, p. ix.
89. Bannan, J., *The Philosophy of Merleau-Ponty*, New York, Harcourt Brace, & World, 1967, p. 46.
90. Ibid., p. 47.
91. Merleau-Ponty quoted by Bannan, Ibid., p. 47.
92. Compare Faust's words: Am enden hangen wir ab von creaturen wir machten. (In the end, we are dependent upon the creatures we have made.)
93. See the Allegory of the Cave, *Republic*, VII, 514a-521b.
94. *Phenomenology of Perception*, p. 207.
95. Ibid., p. 208.
96. Ibid., p. 209.
97. Ibid.
98. Ibid., p. 210.
99. Ibid., p. 211.
100. Ibid., p. 212.
101. Ibid., p. 211.
102. Ibid., p. 214.
103. Ibid., p. 215.
104. Ibid., pp. 215-16.
105. Ibid., p. 216.
106. Ibid., p. 219, also *Sense and Nonsense*, p. 53.
107. Ibid., p. 216.
108. Ibid., p. 222.
109. Ibid., p. 225.
110. Ibid.

111. Ibid.
112. Ibid., p. 223.
113. Ibid.
114. Ibid.
115. Ibid.
116. Ibid., p. 222.
117. Ibid., p. 226.
118. Ibid., p. 225.
119. Ibid., p. 227 and also *Sense and Non-Sense*, p. 50.
120. Ibid.
121. Ibid., p. 229.
122. Ibid.
123. Ibid.
124. Ibid.
125. Ibid., p. 230.
126. Ibid.
127. Ibid., p. 232.
128. Ibid., pp. 232-3.
129. Ibid., p. 233.
130. Ibid.
131. Ibid.
132. Ibid., footnote 1, p. 234.
133. Ibid., p. 235.
134. Ibid.
135. Ibid.
136. Ibid.
137. Ibid, p. 210.
138. Ibid., p. 235.
139. Ibid., pp. 235-6.
140. Ibid., p. 236.
141. Ibid., p. 237.
142. Harry Helson, for example, states that "colors are seldom, if ever experienced as indifferent or neutral... We must therefore seriously consider affective tone as a dimension of color, especially since it has been shown capable of quantitative determination". Helson, Harry, *The Theoretical Foundations of Psychology*, New York, Van Nostrand, 1951, p. 359.
143. Hartshorne, loc. cit., pp. 244-5.
144. Jordan, loc. cit., p. 52.

Southern Illinois University,
Edwardsville, Illinois.

Peirce's First Series of Papers in Logic

by Richard W. Miller

In 1898 Charles Sanders Peirce was to write "Why should there be three principles of reasoning, and what have they to do with one another? This question, which was connected with the other parts of my schedule of philosophic inquiry that need not be detailed, now (in 1867) came to the front."(4.3) This may suggest that in 1898 Peirce remembered that he had been writing the five papers he presented to the American Academy of Arts and Sciences according to some plan. These papers, in their order of presentation, were:

1. "On an Improvement in Boole's Calculus of Logic" Presented March 12, 1867. (1-250-261; 3.1-19)
2. "On the Natural Classification of Arguments" Presented April 9, 1867. (2-261-287; 2.461-516)
3. "On a New List of Categories" Presented May 14, 1867. (3-287-298); 1.545-559)
4. "Upon the Logic of Mathematics" Presented September 10, 1867. (4-402-412; 3.20-44)
5. "Upon Logical Comprehension and Extension" Presented November 13, 1867. (5-416-432; 2.391-426)(1)

It will be the central thesis of this paper to describe what at least a part of the unity Peirce may have intended this series of papers to have was. No great familiarity with the individual papers of this series will be presupposed, although all five papers are reprinted, with minor changes from the original, in the *Collected Papers.* Strangely, although each of the papers seems to have come under study by one Peirce scholar or another, it is seldom that all five have been considered at one time by any single individual.(2) The issue, then, is whether the papers are not treated as a series because they have not been dealt with as such or whether they have not been accepted as a series because of internal evidence to the contrary.

For the most part the more complicated logical and mathematical operations which Peirce included in his papers will be omitted here. While technically interesting, they are not of general significance. It will suffice to say that they can be performed in accordance with normal logical and mathematical procedures.

These papers were presented in March, April, May, September, and November of 1867 to the American Academy of Arts and Sciences. (PAAAS, pp. 249, 261, 287, 402, and 416). They were printed in the *Proceedings* of the Academy in 1868. However, collective reprints of the first three papers seem to have been circulated in December of 1867. This would have been after all five papers had been presented to the Academy; but the reprint of the last two appears not to have been circulated until some time in 1868.(3)

The question of the significance of the division of the papers in the collective reprints is one that will be considered later.

Because he had so recently been elected to the Academy, in January of 1867 (PAAAS, p. 243), Peirce must have been very cautious about the kind of material that he presented to the Academy. He had already given a series of Lowell Institute Lectures and, probably, before that, of University Lectures at Harvard, all on the logic of science. Nevertheless he undoubtedly would have felt compelled to demonstrate his competence in logic to the Academy itself. Given the nature of such a society, Peirce should have been able to presuppose a rather extensive background in both the humanities and the sciences of his average listener or reader. Thus it may be reasonably assumed that Peirce would not have felt the need to include the bulk of traditional logic in these papers, for his audience was probably already familiar with it. German Idealism was the reigning philosophical school in much of America throughout most of the nineteenth century,(4) so Peirce may well have felt that it was proper to assume a fairly detailed knowledge of the writings of Kant, Hegel, and the other major Idealists. Similarly, the "epoch making" work of Boole(5) which had started with his *The Mathematical Analysis of Logic* in 1847 had had sufficient time to have become part of what could be considered common knowledge within this group. Thus Peirce could have taken it for granted that his audience would not need any significant review of what Boole had done. These five papers, then, might not have constituted a comprehensive logic. In all probability they are what Peirce felt were his own independent contributions to logic, for academies existed to foster original work. Popularization was for such agencies as the Lowell Institute. The reason that Peirce believed that the topic under consideration was noteworthy was that:

> Another reform is coming: it is in the air..(sic) Several logical questions are already under discussion by scientific men. Naturalists are divided into two classes, more . . . upon a logical question than anything else. An eminent mathematician (Boole) has proposed a reform of the most important part of the theory of probabilities on logical grounds. . . . On this account, I believe that there are not *now* many thoughtful men of science who will think that the investigation of the character of scientific reasoning is a needless or unimportant inquiry. (Ms. 340. pp. 1-2)

Thus he deals with an *improvement* in Boole's calculus, a *natural classification* of arguments, a *new list* of categories, a *theory* of the derivability of mathematics from logic, and the *implications* of the logical concepts of breadth and depth. Each of these called for a change in the then-existing patterns of thought about logic and was, as Peirce saw it, necessitated by an analysis of things the way they are.

There are many ways that a series of papers by one author may be constituted. It seems that all of these may be grouped into one of three general types: a series may be a purely chronological sequence of papers presented to a single group or in a single publication; it may be a group of papers written upon the same general topic; or it may be a collection of papers on the same basic topic, but where the interrelations among the papers serve to develop a single, more specific topic. The claim that any collection of papers is a series seems to require the least amount of proof in the first case and the greatest amount in the last case. The least significant claim seems already to have been proven; that is, Peirce did present all five papers to the Academy between March and November of 1867. Of course, there is no external evidence to indicate that they were written in the same order in which they were presented, but we do know that they were presented at (nearly) successive meetings of the Academy. It remains to be seen, then, whether claims to being a series of the second and/or third type can be proven.

All but the *New List* can be seen to deal with logic, in some general way, by their titles alone: "On an Improvement in Boole's Calculus of *Logic*"; "On the Natural Classification of *Arguments*"; "Upon the *Logic* of Mathematics"; and, "Upon *Logical* Comprehension and Extension." In the *New List* Peirce writes:

> I shall now show how the three conceptions of reference to a ground, reference to an object, and reference to an interpretant are the fundamental ones of at least one universal science, that of *logic*.(6)

It is in the last several pages of the *New List* that some hint of the unity of the series may be found. As Murphey shows(7) Peirce was strongly influenced by Kant. This is evidenced by his acceptance of the Kantian proof for the necessity of the categories which unite the manifold of experience.

> This paper is based upon the theory already established, that the function of conceptions is to reduce the manifold of sensuous impressions to unity, and that the validity of a conception consists in the impossibility of reducing the content of consciousness without the introduction of it. (3-287; 1.545-61)

There is some question as to whether Peirce introduced three or five categories in the *New List*. In the paper as it was published the *New List* appears:

> BEING: Quality (Reference to a Ground), Relation (Reference to a Correlate), Representation (Reference to an Interpretant),
> SUBSTANCE: The three intermediate conceptions may be termed accidents. (3-293; 1.555)

That exposition is criticised by Manley Thompson as being inconsistent with all his later statements of his categories: its having five categories while the later statements all have only three.(8) In this criticism Thompson may be reading the list differently than Peirce originally intended it to be read. In an earlier draft of the paper Peirce had used the schema:(9)

Categories	Being Accidents Substance	Quality Relation Representation

It may be, then, that Peirce intended in the published paper that there be three categories. However, Peirce does clearly say that "The *five* conceptions . . . may be termed *categories.*"(10) It may also be noted that, in terms of actual applications in the paper, Peirce considers the three "accidents" almost to the exclusion of BEING and SUBSTANCE. The change between the draft and the published manuscript could well reflect a change in Peirce's view of the categories, but that is not at all certain. In any event, whether there are three or five categories, the role of the categories as they are presented in this series is as a basis for logic. Thus, in a very general way, this collection does deal with a single subject—logic. It seems, therefore, quite easily, but perhaps trivially, proven that these five papers do form a series in each of the less significant senses. What is left to be examined is whether or not these papers have more than this superficial unity.

The logic in which Peirce was trained and the logic that he was using in this series was the "traditional logic" which predated the development and use of either propositional or quantificational logic. In this context the central topic of the papers, if there is one, must then relate to categorical, i.e., syllogistic, arguments. Both the second and third papers deal with arguments in broad scope. However, since the obvious focal point of the *New List* is the categories rather than argument, the hypothesis will here be suggested that the most fruitful starting point will be to assume that *Natural Classification* is the central paper of the series.

There are four basic things shown about arguments in this paper: the general form of an argument (2-265; 2.475); the general form of a proposition (2-264; 2.466); and, of primary importance, the classification of arguments into deduction, induction, and hypothesis. (2-281-287; 2.507-517) If, in fact, this paper is the central one in the series, then all of the others must in some way be subordinate to it.

Peirce had said in his University Lectures at Harvard in 1865 that "logical form is already realized in the symbol itself." (Ms. 340, 7th page) Such being his attitude toward logic, it would naturally follow that, in his offerings to the Academy, he would want to provide some theory of symbols and their interrelations for the analysis of logic which he was presenting. Thus he had to develop something at least comparable to the *New List.* There are three elements in a logical argument, i.e., terms, propositions, and an argument, which Peirce further explains by the use of his categories.(11) In the *New List* Peirce has introduced a theory of signs in order to more fully explain the nature of a formal system of symbols. "The objects of the understanding, considered as representations, are symbols, that is, signs which are at least potentially general." (3-294-295; 1.559) But, "all symbols, indeed, are in one sense relative to the understanding...." (3-295; 1.559) Thus, the three elements in logical arguments, as objects of the understanding, must also be signs.

> There would be a general division of symbols, common to all these sciences (formal grammar, formal logic, formal rhetoric); namely, into,
> 1°: Symbols which directly determine only their *grounds* or imputed qualities, and are thus but sums of marks or terms;
> 2°: Symbols which also independently determine their *objects* by means of other term or terms, and thus, expressing their own objective validity, become capable of truth or falsehood, that is, are *propositions*; and
> 3°: Symbols which also independently determine their *interpretants*, and thus, the minds to which they appeal, by premising a proposition or propositions which such a mind is to admit. These are *arguments.* (3-295-296; 1.559)

The terms of the second paper are thus explained in the third as signs which have reference only to their grounds. "... A pure abstraction, reference to which constitutes a *quality* or general attribute, may be termed a ground." (3-291; 1.551) The propositions of *Natural Classification* are signs which employ terms to determine their objects and are either true or false. Its arguments are signs which determine their interpretants and which also use Propositions to add to their import. Thus the elements of an argument presented in *Natural Classification* are made more explicit by the *New List.*

The crucial part of the second paper was the natural classification of arguments into deduction, induction, and hypothesis. Although Peirce does not dwell upon the relation of the categories to these forms of argument, he does give a slight hint of how they might be treated. There is actually no reason to limit the application of these three categories to the traditional usages of 'term', 'proposition', and 'argument', that is, the relation between an argument and a proposition. In his explanation of the kinds of argument, relative to the categories, Peirce seems to indicate that the premises and conclusion might just as well be considered to be terms of a proposition which is the unit we would usually consider to be an argument. This would be a conditional proposition, but Peirce, in *Natural Classification* (2-281n; 2.508n1), indicates that he is accepting the "well-known opinion" that hypothetical propositions can easily be reduced to categorical ones. "In deductive argument, the conclusion is represented by the premises as by a general sign (i.e., a symbol) under which it is contained. In hypotheses, something like the conclusion is proved, that is, the premises form a likeness (i.e., an icon) of the conclusion." (3-296-297; 1.559) However the case is slightly different with induction where "... the premises are an index of the conclusion."(12) Thus the three natural classes of arguments may be differentiated by the difference in the relationships of the signs which form the premises and conclusion.

The hypothesis that *Natural Classification* is the central one gives rise to a question: If it is central, why isn't it presented first? *Boole*, while technically interesting, does not, at first reading, even seem to be related to *Natural*

Classification, *New List*, or *Logical C & E.* Why, then, did Peirce introduce the series with it? Again there seems to be no clear reason stated in the papers themselves. However, a plausible explanation may be made. Remembering that *Boole* was Peirce's first paper as a new member of the Academy, we can imagine that he would have wanted it to serve two functions other than its normal purpose of presenting the material it contains. As a new member Peirce surely would have wanted to present the most clearly proven of his ideas about logic. If this were given as a definite and conclusive proof, he would thereby have presented his professional credentials, so to speak, proving his competence to speak on logic. Each of the other articles(13) seems to rest upon more hypothetical ground; i.e., alternative approaches to the subject matter seem possible in a more complete fashion in them than in *Boole.* Thus, Peirce may have placed his "Improvement in Boole's Calculus" first as his most impressive work. It may also have been the case that he realized the value of relating his own work to other recent developments in logic, thereby making his work fit into a logical evolution, rather than having it appear to be completely novel. None of the other papers would seem to fit this dual criterion of surety and immediacy of interest in general as well as does *Boole.*

The question of the manner in which the final three papers are related to the first, then, is the remaining task. *Boole* develops what Peirce felt was a sufficient means for calculating probabilities. Peirce does not, in this paper, make any claim that this calculus can determine anything about the probability of arguments, or the conclusions of arguments given the probability of the premises, but such a claim does appear implicit in his reformulation of his Example 1. (1-256; 3.15) In this example he symbolizes the hypothetical syllogism:

> If a noise is heard, an explosion always takes place:
> If a match is applied to a barrel of gunpowder, a noise is heard;
> (Therefore): If a match is applied to a barrel of gunpowder, an explosion always takes place.

Here he provides a means of expressing the syllogism in Boolean form. Thus, if the claim of the efficacy of the calculus for figuring probabilities is accepted, he has shown the means of calculating the probability of the conclusion from the probability of the premises. The claim of the expressibility of syllogism in the Boolean calculus is not restricted to hypothetical ones. In his University Lectures he had shown how propositions could be expressed, even when categorical, in the Boolean calculus, and, although the specific tools of formulation changed between that presentation and *Boole*, essentially the same technique could yield a formal Boolean statement of the categorical syllogism. (See 1:253; 3.9; Ms. 342, pp. 18-19) He would have undoubtedly still held that "... in this calculus any number of different equations can be combined into one equation which shall express all the facts contained in the several equations." (Ms. 344, page numbered "Boole 11")

What, then, is the relation of this probability theory of the Boolean calculus to the rest of the series? *Natural Classification*, the paper that is hypothesized to be central, states that "Every inference involves the judgment that, if such propositions as the premises are true, then a proposition related to them as the conclusion is, must be, or is likely to be, true." (2-261; 2.461) This phrase 'is likely to be true' may indicate that probability must be taken into account in determining the strength of any conclusion. However, the traditionally valid figures and moods of the syllogism need not, in Peirce's view, yield necessary conclusions. "It is obvious that the explaining syllogism of an induction or hypothesis may be of any mood or figure." (2-285; 2.512) Thus some means of calculating the probability of premises from terms and of arguments from premises is required by Peirce's "Natural Classification of Arguments." This means of calculation is provided by Peirce's improved Boolean calculus.

Logic of Math is somewhat more problematic than are its predecessors, for about 1905 Peirce was to write:

> I may mention that in a paper of mine published in 1867, I attempted a definition of cardinal numbers considered as multitudinal. But although I received some complimentary letters about that paper at the time, it is now utterly unintelligible to me, and is, I trust, by far the worst I ever published. (4.333)

Also, in 1887, he wrote that it was poor, "as I was then too much given to formal logic to understand mathematical reasoning."(14) If, later in life, Peirce found it so difficult to make sense of this paper, it would seem that there is little hope of fitting it into the pattern set by the other papers in any vital way.

The *Logic of Math* is explicitly tied to the *New List* by Peirce's statement at the end of the fourth paper:

> These considerations, together with those advanced on page 293 (sec. 12) of this volume, will, I hope, put the relations of logic and arithmetic in a somewhat clearer light than heretofore. (4-412; 3.44)

The passage in *New List* to which he refers reads:

> This passage from the many to the one is numerical. The conception of a *third* is that of an object which is so related to two others, that one of these must be related to the other in the same way in .... which the third is related to that other. Now this coincides with the conception of an interpretant. An *other* is plainly equivalent to a *correlate.* The conception differs from that of other, in implying the possibility of a third. In the same way, the conception of *self* implies the possibility of an *other.* The *Ground* is the self abstracted from the concreteness which implies the possibility of another. (3-293; 1.555)

When read with an attempt to see a mathematical meaning, this passage seems truly unintelligible. How ought one to interpret 'third', '*other*', and 'other'? If they were read simply as numbers, the passage could be true only if the relation between them were something specific, such as 'greater than', in which case the 'third' would be the largest of the three, or 'less than', in which case it would be the smallest. Obviously, many other relations of the same type could be imagined, some of which would apply to any triad of numbers, some of which would apply only to triads with special properties. In any event, such interpretations could hardly be looked upon as putting "the relations of logic and arithmetic in a somewhat clearer light." Interpretations which try to put this passage in perspective of the *New List* seem to dissociate it almost totally from mathematics. Exactly what Peirce intended by his reference to the *New List*, then, must wait for some other student of Peirce's thought to decide.

It is also explicitly stated that Peirce plans to use the logical operations which he had introduced in *Boole.* (4-402; 3.20) His objective in writing this paper was "to show that there are some general propositions from which the truths of mathematics follow syllogistically...." (4-402: 3.20) The general propositions from which Peirce believed that he could derive mathematics were the definitions of the logical and arithmetic operators and concepts which he had introduced in *Boole.*(15) The fourteen theorems and six corollaries which Peirce then states (4-403-410; 3.22-41) are, for the most part, not proven. The proofs of most of them are omitted "on account of their ease and want of interest. (4-403; 3.22) The omitted proofs, had they been there, would, however, have been syllogistic in form, thus drawing upon *Natural Classification* and the traditional logic upon which it was based. Peirce did explain that "In order to prevent misapprehension, I will remark that I do not undertake to demonstrate the principles of logic themselves. Indeed, as I have shown in a previous paper, (probably 2-261-262; 2.461-466) these principles considered as speculative truths are absolutely empty and indistinguishable. But what has been proved is the *maxim* of logical procedure, a certain system of signs being given." (4-409; 3.41) In addition this paper might be seen as an addition to the cursory section of the mathematical syllogism in *Natural Classification.* (2-281; 2.507) In that section Peirce had given one example of what he termed "a kind of argument very common in mathematics," (2-281; 2.507) and had altered its form so as to fit it into proper syllogistic form. It might have been that Peirce came to feel that the single example with no exposition was not really a sufficient treatment of mathematical reasoning. If that were the case, he could well have decided, during the four months between the presentations of *New List* and *Logic of Math,* to deliver an additional paper more completely covering that area. Thus this paper was intended to explain, in some fashion, how logic provides a basis for mathematics.

In *Natural Classification* Peirce wrote that:

> Every conclusion may be regarded as a statement substituted for either of its premises, the substitution being justified by the other premises. (2-263; 2.470)

and that

> Every term has two powers or significations, according as it is subject or predicate. The former, which will here be termed its *breadth*, comprises the objects to which it is applied; while the latter, which will here be termed its *depth*, comprises the characters which are attributed to every one of the objects to which it can be applied. This breadth and depth must not be confounded with logical extension and comprehension, *as these terms are usually taken.*(16)
>
> Every substitution of one proposition for another must consist in the substitution of term for term. Such substitutions can be justified only so far as the first term represents what is represented by the second. Hence the only possible substitutions are—
>
> 1st. The substitution for a term fulfilling the function of a subject of another whose breadth is included in that of the former.
>
> 2nd. The substitution for a term fulfilling the function of a predicate of another whose depth is included in that of the former. (2-264; 2.473-474)

Thus inference is justified as a form of substitution according to breadth and depth. However Peirce did indicate that the use of breadth and depth in this context was not the usual one. Since he had written how he was planning to use the terms immediately before his emphasis of the fact that the terms as he used them were not to be "confounded with logical comprehension and extension, as these terms are usually taken," one might wonder why he felt it necessary to mention that his use of the terms was non-standard. This mention seems to indicate that he had plans to further explicate these terms and his use of them at a later point in his papers to the Academy. The entire fifth paper, then, is devoted to a history (thereby showing how these terms are "usually taken") and development of the terms 'breadth' and 'depth'. Thus it seems proper to assume that *Logical C & E* is, in fact, that explication of the concepts of breadth and depth which is required for an understanding of substitution, which is the foundation of inference.

In *Logical C & E* Peirce seems to offer an amplified explanation of the nature of logic which he had introduced in *New List.*

> Logic treats of the reference of symbols in general to their object. A symbol,in its reference to its object, has a triple reference:—
>
> 1st., Its direct reference to its object, or the real things which it represents;
>
> 2nd., Its reference to its ground through its object, or the common characters of those objects;
>
> 3rd., Its reference to its interpretant through its object, or all the facts known about its object.
>
> What are thus referred to, so far as they are known are:—
>
> 1st., The informed *breadth* of the symbol;
>
> 2nd., The informed *depth* of the symbol;
>
> 3rd., The sum of synthetical propositions in which the symbol is

> subject or predicate or the *information* concerning the symbol. (5-429; 2.418)

Thus, at least in relation to symbols, the informed breadth of the symbol seems to correspond exactly to the accident (or category, if it is one) of quality (cf. 3-295; 1.559); the informed depth to the accident of relation (cf. 2-296; 1.559); and the information concerning the symbol to the accident of representation (cf. 2.296; 1.559).

In a letter to the editor of the *Monist* (G-1905-3) Peirce later wrote about his early conception that substitution was the basis of argument, quoting several times from *Natural Classification.* This would seem to indicate that any emphasis placed here on that function would not be inappropriate. There he also indicated that he considered it "*essential* to an argument that it should appeal to the interpreting mind to judge of it independently."(17) In that way he also clarified his position, only hinted at in this series, that the categories as explicated in *Logical C & E* were to serve as an epistemological, rather than a metaphysical, basis for argument.

In addition to the clarification of the meanings of the categories and the terms 'breadth' and 'depth' as they relate to inferential substitution, Peirce used *Logical C & E* to provide a means of distinguishing the three natural classes of arguments by non-formal means, by the analysis of the variation of breadth, depth, and information within each of the kinds of arguments.

> In the case of deductive reasoning it would be easy to show, were it necessary, that there is only an increase of the extensive distinctness of the major, and of the comprehensive distinctness of the minor, without any change in information....
>
> Induction requires more attention. Let us take the following example:
>
> S', S", S" ', and S^{iv} have been taken at random from among the M's;
> S', S", S" ', and S^{iv} are P:
> Therefore any M is P.
>
> We have here, usually, an increase of information. M receives an increase of depth, P of breadth. There is, however, a difference between these two increases. A new predicate is actually added to M; one which may, it is true, have been covertly predicated of it before, but which is now actually brought to light. On the other hand, P is not *yet* found to apply to anything but S', S", S" ', and S^{iv}, but only to apply to whatever else hereafter be found to be contained under M. The induction itself does not make known any such thing. Now take the following example of hypothesis:—
>
> M is, for instance, P', P", P" ', and P^{iv};
> S is P', P". P" ', and P^{iv};
> Therefore S is all that M is.

> Here again there is an increase of information, if we suppose the premises to represent the state of information before the inferences. S receives an addition to its depth; but only a potential one, since there is nothing to show that the M's have any common characters besides P', P", P"', and P^{iv}. M, on the other hand, receives an actual increase of breadth in S, although, perhaps, only a *doubtful* one. There is therefore, this important difference between induction and hypothesis, that the former potentially increases the breadth of one term, and actually increases the depth of another, while the latter potentially increases the depth of one term, and actually increases the breadth of another. (5-431-432; 2.423-424)

Thus Peirce has provided fuller explanations both of the categories and of the three natural classes of argument, doing so in each case in terms of what he defined as breadth and depth. *Logical C & E* must, therefore, have been necessary for the explication of the terms 'breadth' and 'depth'.

The question of the significance of the division of the reprints of the series still wants consideration. If, as Professor Fisch's supplement to Burks's bibliography indicates, the reprint of the first three papers was not circulated until December of 1867, i.e., after the other two papers had been presented to the Academy, it would seem that Peirce must have had some reason for so dividing the papers. There may be, however, suitable reasons for the division's having been made other than Peirce's intention to separate the first three papers from the last two. In 1867 Chauncey Wright was the recording secretary of the Academy. (PAAAS-133, 326) During this period he was also employed as a computer for the *Nautical Almanac.*(18) The papers presented by Peirce during the spring of 1867 would have been given to Wright for preparation for the *Proceedings*, probably during the summer of 1867. Since the reprints were available to Peirce by December, Wright must have had them ready for the printer somewhat before that. Since the linotype machine was not patented until 1884, and was not commercially used until 1886,(19) the copy would have had to been submitted to the printer for hand setting of the type quite a while before the December distribution of the papers. In all probability, Wright would have had all the papers from the previous spring to the printer by early fall, for:

> The computations which, for many years, he (Wright) made for the Nautical Almanac. —as not demanding from him more than two or three hours' work a day, and making no draughts on his thinking power,— furnished in one sense an occupation as favorable as could be imagined for a mind that desired simply leisure for speculation; but, unfortunately, the temptation to him was irresistible to crowd the work into the last two or three months of the year within which it was to be finished, and to try to reduce himself during this period to the insensibility of a computing machine, working from noon until two or three o'clock in the morning.(20)

During the fall of 1867, then, Wright would have had no time to prepare the later papers for the printer. In January of 1868 Wright, having finished for another year his calculations for the *Nautical Almanac*, would have again had time to prepare papers for the printer. The elapsed time, while Wright did his year's work, would have then accounted for the time gap between the appearance of the two collective reprints of Peirce's papers. Further indication that Peirce wanted those who read the first reprint also to read the second one can be found upon the last recto page of the first reprint. Here Peirce indicated:

> Students of logic (and persons who will study it) will be supplied, upon application to the author by letter, with copies of these papers and of others which are to follow. The latter will be sent only to those who express a wish to make use of them.
>
> CHARLES S. PEIRCE
> 2 Arrow St.
> Cambridge,
> Massachusetts.(21)

Thus it seems that there may well be no significance at all in the separation of the reprints of the first three and the last two papers.

In retrospect, the hypothesis of this paper, that *Natural Classification* is the central paper of the series, does appear to have been fruitful. The Boolean calculus, as it was modified by Peirce, can be used to determine the probability of the conclusion of an argument from the probability of the premises. In that sense, *Boole* is subordinate to *Natural Classification. New List* clearly expands the underlying theory of argument found in its predecessor in terms of the new list of categories. The question of what exactly Peirce intended to designate as his new list does not materially affect the relationship between these two papers, for, in practice in this pair of papers, he only uses the three conceptions which he calls 'accidents' as categories. He uses part of *New List* to clarify the three kinds of arguments, the classification of which was the primary function of *Natural Classification. Logic of Math* seems only to have significance as an explanation of the section on mathematical reasoning in the central paper. *Logical C & E* expands and clarifies the means by which inferences can be made by substitution. It does seem clear that *Natural Classification* is related in a positive manner to each of the other articles. *Boole* seems only to be directly connected with the second and fourth papers. *New List* has its primary ties to the second and fifth papers. *Logic of Math* is surely only related to the first two (despite its explicit reference relation to *New List*). *Logical C & E* has direct connections only to the second and third papers. Therefore none of the rest of the series seems capable of serving as the focal point of the series. It must be, then, that Peirce's primary goal in preparing this series of papers was to propose the tripartite division of logic into deduction, induction, and hypothesis. This seems important for the study of Peirce's philosophy, for it indicates that the topic which was most central to his earliest series of papers upon logic was the same as that that was to be a major strand running through the remaining forty-eight years of philosophical endeavor —the analysis of deductive, inductive, and hypothetical reasoning.

The ordering of this series cannot be so clearly shown to be necessary. For reasons explained earlier(22) it seems clear why Peirce presented *Boole* first, even though it is not the central paper of the series. *Logic of Math*, because it draws heavily upon the first two papers, must come after them. In addition, it gives the appearance of being an afterthought: it seems to have come to Peirce that he needed something like this after the basic topics of the series were determined. Also there does not seem to be anything in the unpublished Peirce Papers that deals with the material of this paper at an earlier date than this paper. This is not the case with the other four papers. That is to say, it would appear that the general ideas relating to all but the fourth paper of this series had been under consideration by Peirce for a significant period of time prior to the presentation of the first paper to the Academy. Thus *Logic of Math* appears to have been written during the summer of 1867 upon Peirce's realization that his treatment of mathematical reasoning in *Natural Classification* had been too cavalier. *Logical C & E*, as the promised clarification of 'breadth' and 'depth', would naturally need to come at a point later than either the second or third paper.

Thus, if one were to approach these five papers knowing that they were published in March, April, May, September, and November of the same year to the Academy and that they were all written by the same man, but not knowing in what order they were presented, it would seem probable that one would order them:

1st: Natural Classification [2]
2nd: New List [3]
3rd: Logical C & E [5]
4th: Boole [1]
5th: Logic of Math [4]

If one were given the additional information that the author was a new member of the Academy, that his primary academic and professional field was not logic, and that the last two papers presented might have been after-thoughts, coming after reflection upon the first three papers and a realization that some of the things mentioned in the first three papers might need further explanation, then it seems likely that one would order the papers:

1st: Boole		Boole
2nd: Natural Classification		Natural Classification
3rd: New List	OR	New List
4th: Logic of Math		Logical C & E
5th: Logical C & E		Logic of Math

Thus, with that small bit of additional information, the order in which the five papers occurred is quite understandable.

Therefore it seems that the criteria for being a series in the most significant sense are also met: all five papers develop the single topic of the classification of logical arguments into deduction, induction, and hypothesis.

NOTES

In this paper some works are cited with sufficient frequency to require an abbreviated notation which is incorporated into the body of the paper:

—Burks, A.W., C. Hartshorne, and P. Weiss (eds.). *Collected Papers of Charles Sanders Peirce.* 8 vols. The Belknap Press of the Harvard University Press: Cambridge. 1931-1958.

These will be referred to in the manner common to Peirce scholars: volume and paragraph numbers will be indicated by two Roman numbers separated by a period, the number preceding the period indicating the volume, the one following it indicating the paragraph. Thus 4.456 refers to volume 4, paragraph 456.

The numbers in Burks's bibliography in volume 8 of the *Collected Papers* will be used when referring to items found therein. Items in the supplement to his bibliography (found in Moore, E.C. and R.S. Robin. *Studies in the Philosophy of Charles Sanders Peirce,* Second Series. University of Massachusetts Press: Amherst. 1964.) will be referred to in the same manner with (sup.) following the number.

Unpublished manuscripts in the Peirce Papers in Houghton Library, Harvard University, are referred to by the numbers they bear in Richard S. Robin's arrangement of them.

Page references to the five articles by Peirce which are the basis for this paper are given by indicating which of the papers is referred to, the order of the papers 1 to 5 being indicated on page 72 (above), followed by a dash and then the page reference. All five papers are found in Proceedings of the *American Academy of Arts and Sciences.* Volume 7. 1867-68. References to material in this volume, not in the five papers, is made by PAAAS followed by the page number. When there is a reference to a passage found both in the *Collected Papers* and in the *Proceedings*, the *Proceedings* number will come first, followed by the *Collected Papers* number. Thus, (2-263; 2.470) refers to the second paper of the series, page 263 of volume 7 of the *Proceedings* and to volume 2 of the *Collected Papers*, paragraph 470.

*****

1. These five papers will be referred to by the following abbreviated titles:
"On an Improvement in Boole's Calculus of Logic" will be *Boole.*
"On the Natural Classification of Arguments" will be *Natural Classification.*
"On a New List of Categories" will be *New List.*
"Upon the Logic of Mathematics" will be *Logic of Math.*
"Upon Logical Comprehension and Extension" will be *Logical C & E.*

2. The reason for this failure to consider all five papers as a unit may result from the editing of the *Collected Papers.* With this series their lack of regard for chronology of Peirce's writings is especially troublesome. Of the papers in this series the first and fourth are found in volume 3 of the *Collected Papers*; the second and fifth are in volume 2; the third is in the first volume. Not only are they widely separated, but the notes at the beginning of each of them gives only the location where the original is to be found, but no indication that other papers are to be found in the same location. Burks's bibliography shows that there are five papers in the group, but it was not published until 1958, twenty-five years after the last volume in which these papers is found was published. In addition the *Collected Papers* reprint of these articles insists upon using the spelling 'premiss' which Peirce adopted later in life. In this paper the spelling 'premise' which was used in the original publication is retained.

3. G-1867-1(a-c) (sup.) says December of 1867, however *The Monist*, Vol. 15, p. 294, indicates that the first reprint was circulated in the summer of 1867. Also see bound copies of these two reprints in the library of Max H. Fisch.

4. Henry Steele Commager. *The American Mind,* (New Haven: Yale University Press, 1961) pp. 87, 92.

5. On which see I.M. Bochenski. *A History of Formal Logic,* (Notre Dame: University of Notre Dame Press, 1961) p. 278.

6. 3-294; 1.559. Emphasis mine in all italics here.

7. Murray Murphey. *The Development of Peirce's Philosophy*, (Cambridge: Harvard University Press, 1961)

8. Manley Thompson. *The Pragmatic Philosophy of C.S. Peirce*, (Chicago: University of Chicago Press, Phoenix Books, 1963) pp. 19-29.

9. Ms. 732. p. 9; Murphey. p. 419.

10. 3-293; 1.555. Former emphasis mine, latter Peirce's.

11. It is well here to recall that, whether or not Peirce intended there to be three or five categories and despite the fact that BEING and SUBSTANCE seem quite clearly to be categories in his presentation, what he actually uses, in most cases, as categories are the three accidents. Thus in general usage, the categories for Peirce seem to have been Quality, Relation, and Representation.

12. 3-297; 1.559. Emphasis mine.

13. With the possible exception of *Logic of Math* which, although a proof, does not seem as fully worked out and, in any case, rests upon the proofs of *Boole* and *Natural Classification.*

14. A letter from Peirce to J.M. Hantz, Professor of Mathematics at Northwestern University, Watertown, Wisconsin, March 29, 1887.

15. These definitions are stated at (4-402-403; 3.21) and (1-250-253; 3.2-7)

16. Italics mine.

17. *The Monist,* Volume 15, April 1905, p. 295.

18. Edward H. Madden. *The Philosophical Writings of Chauncey Wright,* (New York: The Liberal Arts Press, 1958), p. viii.

19. J. N. Kane. *Famous First Facts,* Third Edition. (New York: H. W. Wilson Co., 1964), p. 635.

20. J. B. Thayer (ed.). *The Letters of Chauncey Wright,* (Cambridge, Massachusetts: John Wilson, 1878), pp. 376-377.

21. Bound copy of "Three Papers in Logic," last recto page, in the library of Professor Max H. Fisch.

22. See p. 77, above.

—An expanded version of this paper served as the author's M.A. thesis at the University of Illinois, 1966.

University of Missouri-Rolla,
Rolla, Missouri

Louis Agassiz and the Origin of Species

by David E. Pfeifer

In July of 1858 a man stood before an assembly of scholars and announced, "I am no longer a European." The assembly rose to its feet and greeted the announcement with joyous and repeated cheering.(1) The man who provoked this emotional outburst was Jean Rudolphe Agassiz, better known as Louis Agassiz.

"From his boyhood he looked on the world as if it and he were made for each other, and on the vast diversity of livingthings as if he were there with authority to take mental possession of them all....[Agassiz] was recognized by all as one of those folio copies of mankind...who aim at nothing less than an acquaintance with the whole of animated Nature."(2) Born at Motier, Switzerland, on May 28, 1807, Louis Agassiz received his education at Lausanne, Zurich, Heidelberg, and Munich. Naturalists first became aware of Agassiz' capabilities between 1833 and 1844 when he published a series of outstanding works on fossil fishes. In 1840 he established a finding which in itself would have preserved his name in any list of notable natural scientists. This finding was that the earth was once covered with ice; the Ice Age and the name of Louis Agassiz are forever joined. This titan of a man came to Boston in 1846 to deliver a series of lectures. All that was needed for this trip was a one-way ticket. After his resettlement Agassiz spent his life collecting specimens of every kind of organism, and in the process he gathered a museum of factual material which he systematized into a body of knowledge with almost cosmic dimensions. In addition to his massive intellect, Agassiz had a personality as broad and as deep as his learning. He was loved and respected, a man who stood above the heads of those around him, and when he spoke, people listened. Indeed, he was asked to speak time and time again, being one of the most popular lecturers in Boston.

Late in life, however, one incident came close to relegating Louis Agassiz to the intellectual oblivion of naturalists who propound misdirected theories. This incident arose with the publication of Charles Darwin's *The Origin of Species*, the ensuing controversy lasting until Agassiz' death in 1873. The developmental hypothesis of Darwin's theory of evolution contradicted everything Agassiz taught. The rise of this new theory, coupled with the fact that Louis Agassiz was recognized as *the* authority on natural science in the United States, resulted in Agassiz taking his stand against Darwinism and proclaiming that he would outlive this mania.(3) Max H. Fisch pointedly notes: "No good cause should be without an eminent and vigorous adversary, who prevents it from being ignored. Such an adversary was Louis Agassiz."(4) If Agassiz had not staunchly opposed transmutational evolution, the doctrine may have quietly passed into use without any great furor. It may even have fallen on silence as did the work of Gregor Mendel in 1865, which would have directly supported the Darwinian hypothesis had naturalists been aware of it.(5)

Today Darwinian evolution is rightly accepted as a proven theory, yet the conclusive evidence for transmutational evolution did not begin to appear until the 1930's when the work of Gregor Mendel was rediscovered.(6) Until that time evolution remained a hypothesis, an open question. Scientists accepted Darwin's *The Origin of Species* when it first appeared, with Agassiz being the only capable opposition within natural science. The fact is that even Agassiz' own students —including his own son Alexander— placed themselves within the Darwinian camp. Some claimed that it was Agassiz' arguments against Darwin which finally pushed them into the evolutionary camp. This assertion is only partially correct. In retrospect, it is seen that they reacted to the viewpoint from which Agassiz was arguing; they reacted to his non-scientific theory of special creation and his dogmatic adherence to this theory. These opinions created a distrust in Agassiz' ability as a theorist (his scientific ability was never in doubt), which led to naturalists feeling that Agassiz' entire theory and his opposition to Darwin were very weak.

Before the content of the controversy between Agassiz and Darwinism is discussed, there is need for some background material.(7) As might be expected, when Darwin's work appeared in the United States, two camps arose —those for Darwin and those against Darwin. However, before the publication of *The Origin of Species* the leaders of these camps had already arisen, marked off their distances, and engaged in some preliminary skirmishes.(8) Around these two leaders the respective armies developed.

The leader of Darwin's supporters was Asa Gray, a professor of botany at Harvard College. (Agassiz was professor of zoology at Harvard.) Gray presented the first sympathetic review of Darwin's work to the scientific community in the *American Journal of Science and Arts*, and presented the first friendly discussion of Darwin to the American public in the *Atlantic Monthly*.(9) In addition to these firsts, Gray published many other articles on evolution over the next several years, all of which he then reprinted in book form under the title *Darwiniana* in 1876.(10) When the debates on evolution began in the American Academy of Arts and Sciences, it was Gray who stood up to Agassiz and defended Darwin's position.

> Gray's support was admirably timed. It coincided almost exactly with Agassiz' opposition and did much to counterbalance it. Although conjecture is admittedly unsatisfactory, it is reasonably safe to conclude that given Agassiz and no Gray it would have been very difficult for Darwin to have obtained so ready a hearing in this country. It was Gray who braved Agassiz' opposition and prepared the ground for an ultimate acceptance of evolutionary views.(11)

While Gray was busy writing about evolution and meeting with Agassiz in the American Academy of Arts and Sciences, William Barton Rogers, an American geologist who was instrumental in the founding of the Massachusetts Institute of Technology, was debating with Agassiz in the Boston Society of Natural History. In 1846, Gray helped block the appointment of Henry Rogers, William's brother, to the chair of geology at Harvard, but now that there was a common enemy Rogers and Gray were slightly friendlier to each other.(12)

Chauncey Wright, the third important participant in this controversy in the Darwinian camp, was an American philosopher who is most noted for his work in freeing science from its metaphysical and theological bindings; he is also noted for being a mentor to Charles S. Peirce and William James. "In 1860 Chauncey began teaching natural philosophy, or science, in Louis Agassiz' school for young ladies in Cambridge. At this time, ironically, Wright first read Darwin's *Origin of Species....*"(13) It should also be noted that Wright studied under Asa Gray. In 1870, Wright published his first article in the *North American Review* defending the theory of Darwin; one article a year followed for the next three years.(14) Wright is very important in considering this controversy between Agassiz and Darwinism, for he sought to straighten out some of the theoretical and philosophical difficulties in evolution. Wright's work becomes especially important, when it is realized that Agassiz' strongest criticisms were philosophical in nature.

The leader of the other camp was of course Louis Agassiz. It might be said that he was fighting both ends against the middle. He was not only debating the Darwinians, but the various religious journals of the day were continually attacking his theory of special creation.(15) If this issue is left to the side, there is still much material to be considered. Fortunately, however, Agassiz made two major statements of his criticisms of the theory of evolution; these two articles will be the major stepping stones for this paper. The first article was published in the *American Journal of Science and Arts*, of which Agassiz was an editor, during the height of battle in 1860, and the second article was published posthumously in the *Atlantic Monthly*.(16) This second article was a more reserved appraisal of Darwin and was intended to be the first of a series of papers which would systematically refute Darwin.

In the debates at the American Academy of Arts and Sciences, there were two prominent persons who took their stand beside Agassiz —Francis Bowen and John A. Lowell. Francis Bowen was the Alford Professor of Moral Philosophy at Harvard and had taken his own shot at Darwin.(17) John Lowell also made an attack on Darwin,(18) but more importantly he was the man who in 1846 made it possible for Agassiz to remain in the United States. Indeed, he and other prominent Bostonians did not want to lose such an illustrious scientist.(19) Lowell possessed "a considerable share of the wealth and power of Boston," and "was the leading member of the Harvard Corporation."(20) It seems that Lowell had the political power to put pressure on Gray if he wanted, but he never used it, although his presence in the battle heightened Gray's combativeness, if anything.

The difference between Agassiz' theory of special creation and Darwin's theory of evolution can be easily stated. In fact, Asa Gray has done it so well it is worthwhile to let him relate their views.

> The ordinary and generally-received view [of the origin of species] assumes the independent, specific creation of each kind of plant and animal in a primitive stock, which reproduces its like from generation to generation, and so continues the species. Taking the idea of species from this perennial succession of essentially similar individuals, the chain is logically traceable back to a local origin in a single stock, a

single pair, or a single individual, from which all the individuals composing the species have proceeded by natural generation. Although the similarity of progeny to parent is fundamental in the conception of species, yet the likeness is by no means absolute; all species vary more or less, and some vary remarkably —partly from the influence of altered circumstances, and partly (and more really) from unknown constitutional causes which altered conditions favor rather than originate. But these variations are supposed to be mere oscillations from a normal state, and in Nature to be limited if not transitory; so that the primordial differences between species and species at their beginning have not been effaced, nor largely obscured, by blending through variation. Consequently, whenever two reputed species are found to blend in Nature, through a series of intermediate forms, community of origin is inferred, and all the forms however diverse, are held to belong to one species. Moreover, since bisexuality is the rule in Nature (which is practically carried out, in the long-run, far more generally than has been suspected), and the heritable qualities of two distinct individuals are mingled in the offspring, it is supposed that the general sterility of hybrid progeny interposes an effectual barrier against the blending of the original species by crossing.

From this generally-accepted view the well-known theory of Agassiz and the recent one of Darwin diverge in exactly opposite directions.

That of Agassiz differs fundamentally from the ordinary view only in this, that it discards the idea of a common descent as the real bond of union among individuals of a species, and also the idea of a local origin —supposing instead, that each species originated simultaneously, generally speaking, over the whole geographical area it now occupies or has occupied, and in perhaps as many individuals as it numbered at any subsequent period.

Mr. Darwin, on the other hand, holds the orthodox view of the descent of all the individuals of a species not only from a local birthplace, but from a single ancestor or pair; and that each species has extended and established itself, through natural agencies, wherever it could; so that the actual geographical distribution of any species is by no means a primordial arrangement, but a natural result. He goes further, and this volume [*The Origin of Species*] is a protracted argument intended to prove that the species we recognize have not been independently created, as such, but have descended, like varieties, from other species. Varieties, on this view, are incipient or possible species: species are varieties of a larger growth and a wider and earlier divergence from the parent stock; the difference is one of degree, not of kind.

The ordinary view —rendering unto Caesar the things that are Caesar's— looks to natural agencies for the actual distribution and perpetuation of species, to a supernatural for their origin.

The theory of Agassiz regards the origin of species and their present general distribution over the world as equally primordial, equally supernatural; that of Darwin, as equally derivative, equally natural.

The theory of Agassiz, referring as it does the phenomena both of origin and distribution directly to the Divine will —thus removing the latter with the former out of the domain of inductive science (in which efficient cause is not the first, but the last word)— may be said to be theistic to excess. The contrasted theory is not open to this objection. Studying the facts and phenomena in reference to proximate causes, and endeavoring to trace back the series of cause and effect as far as possible, Darwin's aim and processes are strictly scientific, and his endeavor, whether successful or futile, must be regarded as a legitimate attempt to extend the domain of natural or physical science.(21)

The conflict between Agassiz' theory of special creation and Darwin's theory of development by means of transmutation is not a question concerning the facts of natural science, but is a matter of the interpretation of these facts. Agassiz interpreted the actualities of nature as a representation of the thought and action of God, whereas Darwin felt that natural phenomena could only be explained by the supposition of the modification of species.(22) Here there is a subtle but important distinction. Both Agassiz and Darwin were collectors of factual material concerning natural science, and with this material they could develop a scheme within which all items would be characterized and classified. On this primary classification they would be in basic agreement. Beyond this process of classification Agassiz would not go, for he felt that human knowledge had not reached the point at which it could proceed further than this.(23) The scientific aspect of Agassiz' position is found in his study of the plan, order, and connection within the phenomena of nature. The interpretation of these findings as a manifestation of God's thought is philosophical, speculative, and not an instance of scientific knowledge; this distinction was clear in Agassiz' mind and can be seen in his "Essay on Classification."(24) Darwin on the other hand possessed the same set of facts, and wanted to explain the reason behind the order and connection of natural phenomena. To accomplish this task he made a certain assumption —the transmutation of species— which, he maintained, would explain the phenomena of nature, in that the course of nature demonstrates and manifests the operations of those laws contained in the developmental hypothesis. Both men took a step beyond the established truth of natural science, but where Agassiz is looking up from this truth to a higher order, Darwin is looking behind it in an attempt to explain its nature.

On the whole, naturalists sided with Darwin as possessing the better explanation of natural phenomena, while considering Agassiz' opposition to Darwin unscientific and very poor. However, there is one part of this controversy which is commonly overlooked; this is Agassiz'critique of *The Origin of Species.* As Agassiz was the major opposition to Darwin, it seems that his views should have been discussed by Darwin's defenders. Such a discussion was never written.(25) The major reason is that Agassiz' views on the origin and distribution of species were thoroughly discussed at the various scientific academies before and after the publication of Darwin's work.(26) Since these views were thoroughly discussed and completely rejected, no one felt it necessary to meet Agassiz' criticisms when they did appear. This discussion will therefore be an attempt to present Agassiz' criticisms to determine if they could have been answered, and in what follows, there is an attempted reconstruction of some of the reasons why Agassiz' views were rejected. These reasons will be based primarily upon the content of Agassiz' critique, assuming of course that scientific opinion was on the side of Darwin.

Agassiz was an observer of discussions on natural philosophy in his youth, was conversant with the writings of Plato and Aristotle, and had studied German metaphysics through Hegel.(27) On the other hand, Darwin studied Paley's *Evidences of Christianity*, *Moral Philosophy*, and *Natural Philosophy* at Cambridge University, with his opinion of this kind of study being: "The careful study of these works, without attempting to learn any part by rote, was the only part of the academical course which, as I then felt, and as I still believe, was of the least use to me in the education of my mind."(28) Agassiz was willing to philosophize; this was a constituent of his view of science, whereas Darwin "simply refused to philosophize at all."(29) Based upon this information, it should be relatively clear that Darwin would not be as cognizant of the philosophical implications of this theory as Agassiz. It should also come as no surprise that Agassiz' most telling and pointed criticisms of Darwin are philosophical in nature, not scientific. This information also explains why Darwin was happy to have Chauncey Wright and Asa Gray come to his aid in solving some of these difficulties. Indeed, it will be found that it is Wright and Gray, not Darwin, who answer Agassiz' criticisms.

As in most controversies, some criticisms hit the mark and some miss the mark. Since the emphasis here is upon Agassiz' role in the debate, his criticisms will be presented and then Darwin's or his supporters' answers, if any, will be given to determine if Agassiz did hit the mark. Agassiz made many critical statements; the most general of these will be presented first, after which more specific statements will be considered.

It is first found that Agassiz disagrees with what Darwin has put forth as science. He states that

> the office of science is not to record possibilities; but to ascertain what nature does.... As far as Darwinism deals with mere arguments of possibilities or even probabilities without a basis of fact, it departs from the true scientific method & injures science,as most of the devotees of this new ism have already done.(30)

This statement was taken from a private letter which was written in 1873, and shows Agassiz' more emotional reaction to Darwinism. His published statement of 1874 is more reserved.

> The study of the relations of animals and plants has at all times been one of the principal aims of all the more advanced students of Natural History....Darwin has placed the subject on a different basis from that of all his predecessors, and has brought to the discussion a vast amount of well-arranged information, a convincing cogency of argument, and a

captivating charm of presentation. His doctrine appealed the more powerfully to the scientific world because he maintained it at first not upon metaphysical grounds but upon observation. Indeed, it might be said that he treated his subject according to the best scientific methods, had he not frequently overstepped the boundaries of actual knowledge and allowed his imagination to supply the links which science does not furnish.(31)

For Agassiz "the criterion of a true theory consists in the facility with which it accounts for facts accumulated in the course of long-continued investigations and for which the existing theories afforded no explanation."(32)

If these statements are put together, according to Agassiz, Darwin is making two fundamental errors: One, Darwin is overstepping the boundaries of what is considered as scientific fact; and two, since the facts have been satisfactorily explained, no new theory is really needed as yet. On the other hand, Darwin would charge Agassiz with his own first criticism, while denying the second. Both Agassiz and Darwin are holding different views of the nature of science and the form of theory appropriate to science; the two are simply contrasting their separate views on the subject with neither one demonstrating flaws within the other position.

Both Agassiz and Darwin felt that they possessed the correct view of the nature of science. It is now known that Darwin was correct in his view, but at the time he held this opinion, he did not know why he was right; this is the point where Chauncey Wright became important in philosophically explaining and justifying Darwin's position. Wright states his conclusions as follows.

> The very hope of experimental philosophy, its expectation of constructing the sciences into a true philosophy of nature, is based on the induction, or, if you please, the a priori presumption, that physical causation is universal; that the constitution of nature is written in its actual manifestations, and needs only to be deciphered by experimental and inductive research; that it is not a latent invisible writing, to be brought out by the magic of mental anticipation or metaphysical meditation. Or as Bacon said, it is not by the "anticipation of the mind" but by the "interpretation of nature" that natural philosophy is to be constituted; and that this is to presume that the order of nature is decipherable, or that causation is everywhere manifest or hidden but never absent....(33)

Agassiz could not answer Wright; he could only say that Wright was mistaken. For Agassiz, questions in science were matters of fact; science was deductive and certain. With Wright, science is inductive and fallible; questions in science are matters of possible explanation, not fact. Agassiz and Wright are taking different viewpoints on the same issue with Wright's view being the one that is considered as philosophically and scientifically more justifiable. Indeed, in this issue and in all issues where it is a contrasting of viewpoints, scientific opinion is opposed to the views presented by Agassiz.

Another case of contrasting viewpoints, or an opposition of basic assumptions, is found in the question of the nature of the geological record. Agassiz concluded from his studies that "throughout all geological times each period is characterized by definite specific types, referable to definite orders, constituting definite classes and definite branches, built upon definite plans."(34) Thus, "were the transmutation theory true, the geological record should exhibit an uninterrupted succession of types blending gradually into one another."(35) Darwin held the opposite interpretation, namely, that the geological record does not yield much knowledge and cannot be trusted. "The crust of the earth with its imbedded remains must not be looked at as a well-filled museum, but as a poor collection made at hazard and at rare intervals."(36) With this and analogous statements Darwin has thrown off the task of reconciling his theory with the historical record of organic life as found in the crust of the earth. By taking a position opposite Agassiz, Darwin did not have to meet on geological grounds Agassiz' strongest factual counter-example —the fact that higher (more complex) fish are found in a geological period earlier than low (less complex) fish.(37) Darwin could not answer this query with geological evidence, but could assert that this phenomenon was completely compatible with his theory. Since natural selection determines what is advantageous for an organism within its environment, it is possible for regression to take place as well as the more common progression. The only thing that Darwin needed to be consistent in his theory was the establishment of the hypothesis that earliest life was of a low form; if this hypothesis is granted or proven, Agassiz' counter-example does not hold.

Darwin was forced to admit that he did not have the means of explaining how transmutation was effected,(38) but he did feel that "whatever the cause may be of each slight difference between the offspring and their parents —and a cause for each must exist— we have reason to believe that it is the steady accumulation of beneficial differences which has given rise to all the more important modifications of structure in relation to the habits of the species."(39) Natural selection does not induce mutation; selection is only "the preservation of such variations as arise and are beneficial to the being under its conditions of life."(40) To illustrate this point Darwin discussed selection as found in animals under domestication, and as found in a natural state.

Agassiz attacked this doctrine on three points. First, animals do not transmit lasting gains of the type proposed by Darwin.(41) Here, Agassiz is simply opposing his interpretation of the same information possessed by Darwin. He misses the mark, for he cannot conclusively establish this assertion, but for that matter neither can Darwin establish its opposite. Second, domesticated animals are "never traced back to anything but their own species, nor have artificial varieties, so far as we know, failed to revert to the wild stock when left to themselves."(42) Here, the same applies as in the first case. And third, "Mr. Darwin has failed to establish a connection between the mode of raising domesticated breeds, and the cause or causes to which wild animals owe their specific differences."(43) Agassiz argues that the improvement of animals by man with selective breeding is really a process of control which implies design, i.e., breeding for a purpose. In a state of nature neither control nor design are found. Therefore, even if some kind of selection is granted with regard to the breeding of animals, and Agassiz grants this,(44) Darwin would have to make a connection between this form of selection and the selection supposedly found in a state of nature. The heart of the empirical justification for Darwin's theory lies in the domestication and breeding of animals. Darwin followed this material with the hypothesis that the same advantageous conditions which produce varieties under domestication could arise in nature and bring about the same result. The difficulty which Agassiz is pointing to was not felt by him alone; Asa Gray was fully aware of this missing link in Darwin's theory and noted it in his review of *The Origin of Species.* This difficulty led Gray to consider Darwin's hypothesis as only probable, although probably correct.(45) The problem here, which falls more on Darwin and his defenders than on Agassiz, is that the mechanism by which variations arose was not known. If this information had been known, Darwin would not have had to consider animals under domestication and in a natural state as two separate, yet connected, instances of the same phenomenon; he could have simply considered organisms.

Something was already known of the science of genetics when the evolutionary debate took place. Genetics was accepted as playing a role in the procreation of species, but as Gray's biographer notes: "To have accepted Darwin on genetics would have been quite as obscurantist and antiscientific as to have accepted the first chapter of *Genesis.*"(46) This consideration leads one to believe that the major difficulty of Darwin's theory would remain until the science of genetics was given a more important place within biology and zoology. Historically, this is what happened; neither Darwin, Gray, nor Wright would find the missing link during their lifetimes.

Thus far the controversy between Louis Agassiz and Darwinism has been basically between varying interpretations and viewpoints on the same subjects; each theory is more or less consistent within itself, and when they meet it is found that one is opposite the other. In order to show that one is correct and the other wrong, it would be necessary to step outside both theories and gain more information or present arguments independent of both theories. More information was not at hand, so this avenue was closed, and Agassiz never left the confines of his own theory, so he did not establish that his theory was of more worth. However, there remains the possibility of stepping inside the other's theory and showing that it contains flaws, thereby showing that it cannot be the correct theory. This Agassiz did do; he entered Darwin's theory on Darwin's own grounds and attempted to show that Darwin made some philosophical errors which would harm the underlying assumptions of his theory. These criticisms center around Darwin's conception of species and his notion of classification.

The title of Darwin's work is *The Origin of Species*; this title by itself indicates that Darwin is concerned primarily with this category of classification of organic life. When Agassiz first saw this book with its startling title, he probably sat straight in his chair and stared in amazement. Agassiz would have been startled, for he firmly believed that the study of Species as the basis of a scientific education is a great mistake. It leads us to overrate the value of Species, and to believe that they exist in Nature in some different sense from other groups; as if there were something more real and tangible in Species than in Genera, Families, Orders, Classes, or Branches. The truth is, that to study a vast number of Species without tracing the principles that combine them under more comprehensive groups is only to burden the mind with disconnected facts....(47)

Louis Agassiz spent his life gathering information about organisms and considered a major part of his task their arrangement into a coherent system which separated these organisms into groups which paralleled the way in which they are segregated in nature as closely as possible. For this reason, a species is no more of a special object of study than any one of the other categories. Like other classifications, "while species have no material existence, they yet exist as categories of thought."(48) Classifications are all based on "relations and proportions" and "categories of characters".(49) According to Agassiz, when there arises a question of whether a particular specimen belongs to one species or another, there is no question of the specimen's origin, but only a question of its characteristics and properties. Classification is largely a matter of division and differentiation. If items are separated into different categories, they are different. Therefore, by placing items in different categories, they are different. Therefore, by placing items in different species, genera, etc., one is implicitly asserting that they are not related, which rules out the idea of a common descent. If put in oversimplified terms, Agassiz has before him a vast number of variously characterized items which he sorts, arranges, and catalogues in a system of classification which puts the most similar together in some kind of ascending order of generality.

"Agassiz has comparedhis system of classifying animals to the grouping of the stars; the stem and branches of a tree better illustrates Darwin's idea."(50)

Darwin felt that the system of classification proposed by Agassiz was a mere useless abstraction.(51) On the other hand, he asserted that

all the...difficulties in classification may be explained, ...on the view that the Natural System is founded on descent with modification; —that the characters which naturalists consider as showing true affinity between any two or more species, are those which have been inherited from a common parent, all true classification being genealogical; —that community of descent is the hidden bond which naturalists have been unconcsiously seeking, and not some unknown plan of creation, or the enunciation of general propositions, and the putting together and separating of objects more or less alike.(52)

With reference to Darwin's mode of classification, Agassiz raises two major points. First, "we do the thing before we understand why we do it: speech precedes grammar, reason precedes logic; and so a division of animals into groups, upon an instinctive perception of their differences, has preceded all our scientific creeds and doctrines."(53) Darwin would deny Agassiz' method of classification, yet if there were no such classification, Darwin could not have formulated his theory. Unless there are some fairly well defined categories to start with, it is impossible to assert that an individual or a species deviates from a category. Darwin is working within the Agassizian type of classification attempting to set up a new form of classification which overthrows the older form. To establish this new classification Darwin would have had to begin with some species and then trace its development until it branched out, via natural selection and transmutation, into a clearly defined family tree. Darwin lacked the means of accomplishing this task.

Agassiz' second point is that "if species do not exist at all, as the supporters of transmutation theory maintain, how can they vary? and if individuals alone exist, how can the differences which may be observed among them prove the variability of species? "(55)

> Unless Darwin and his followers succeed in showing that the struggle for life tends to something beyond favoring the existence of certain individuals over that of other individuals, they will soon find out that they are following a shadow. The assertion of Darwin, which has crept into the title of his work, is, that favored *races* are preserved, while all his facts go only to substantiate the assertion, that favored *individuals* have a better chance in the struggle for life than others. But who has ever overlooked the fact that myriads of individuals of every species constantly die before coming to maturity? What ought to be shown, if the transmutation theory is to stand, is that these favored individuals diverge from their specific type, and neither Darwin nor any body [*sic*] else has furnished a single fact to show that they go on diverging.(56)

> The great defect in Darwin's treatment of the subject of species lies in the total absence of any statement respecting the features that constitute individuality. Surely, if individuals may vary within the limits assumed by Darwin, he is bound first to show that individuality does not consist of a sum of hereditary characteristics, combined with variable elements, not necessarily transmitted in their integrity, but only of variable elements.(57)

These quotations present Agassiz' best statement of the theoretical flaws that are found in Darwin's theory. Here, three things are asked for: one, a definition of species; two, a definition of individuality; and three, the relation between individuals and species. Unless individuality is defined, i.e., it is asserted which characteristics belong to an individual as a member of a species and which characteristics are peculiar to that individual, one cannot make any statements about the individual deviating from its specific type, its species. And even if this individual does vary, one must somehow show the manner in which the change in this individual brings about a new or different species.

These questions were not answered by either Darwin or Gray. Although

Darwin's theory of evolution was generally accepted, his "contemporaries intimated that J.L.R. Agassiz' theory of species was real, fundamental and absolute, while Darwin's own view of the species classification was conventional, arbitrary and relative. Darwin did not like this description of his view but was unable to put his finger on what was wrong with it. Consequently, he was quite pleased when [Chauncey] Wright came to his aid."(58) Wright, as a philosopher interested in the logic of science, was keenly aware of the difficulties pointed out by Agassiz, and he did answer Agassiz' queries.

According to Wright,(59) there is a difference between a real or fundamental class or species and a conventional or arbitrary class. A class is real when the individuals composing the class have a cluster of characteristics in common, whereas a class is conventional when the individuals composing the class have only the characteristics in common which formed the basis of classification. With this distinction in hand, Wright can assert: One, a species is composed of a group of individuals which manifest a cluster of like characteristics; two, an individual is a member of a species when it possesses this cluster of like characteristics; and three, a new species can arise when a group of individuals arises with a new cluster of like characteristics. (This group could arise from a single individual.) What Wright asserts is perfectly compatible with Darwin's theory, but would cause Agassiz to frown. Agassiz would frown for he maintains the fixity of species which appears incompatible with what Wright has presented. Wright would answer Agassiz by asserting that animals do not exhibit a fixity of characteristics, but a distinctness of characteristics. Fixity implies duration through all time, whereas distinctness only implies duration through an indefinite period of time. Thus, Wright, and Darwin with him, would argue that this idea of species explains, or rather allows for, the creation and destruction of species; one species may exist without change for all time, while another species may only exist for several generations. The concept of species offered by Wright is appealing for it encompasses both the Agassizian theory of species and the Darwinian. The only question left remaining by Wright's analysis of species is how an individual, who could procreate sufficient numbers to generate a class of individuals that should be termed a species, comes to possess a new cluster of characteristics. Wright's answer is that there are physiological and physical causes which can be found in the laws of inheritance and the action of natural selection.(60) What these causes are he cannot say.

It has been found that all of Agassiz' criticisms are simply different viewpoints or assumptions when compared to Darwinism, or they are such that they could be met by Darwin's supporters, if they had taken the time to address themselves to these views. There is, however, one final point which should be mentioned, and this is that Agassiz felt evolution would remove the majesty of God from the realm of science —Agassiz' belief in the divine creation and plan of nature would be lost. In answer to Agassiz, Asa Gray, who was a theist and held something similar to Agassiz in this respect, presents a parable which, had Agassiz paid attention to it, would have salved his religious feelings.

Recall a woman of a past generation and show her a web of cloth; ask her how it was made, and she will say that the wool or cotton was carded, spun, and woven by hand. When you tell her it was not made by manual labor, that probably no hand has touched the materials throughout the process, it is possible that she might at first regard your statement as tantamount to the assertion that the cloth was made without design. If she did, she would not credit your statement. If you patiently explained to her the theory of carding-machines, spinning-jennies, and power-looms, would her reception of your explanation weaken her conviction that the cloth was the result of design? It is certain that she would believe in design as firmly as before, and that this belief would be attended by a higher conception and reverent admiration of a wisdom, skill, and power greatly beyond anything she had previously conceived possible.(61)

Agassiz looked upon classification as a matter of logical division, a matter of sorting out natural phenomena as it is found; Darwin looked upon classification as a matter of tracing the history of natural phenomena. For Agassiz, natural phenomena was a given which was to be understood; for Darwin, natural phenomena was that which must be explained. Agassiz was given nature by God as a complete system and could develop an absolute doctrine; Darwin found a dynamic nature wherein he sought to explain the changes which took place and could only develop a fallible theory.

The controversy between Agassiz and Darwinism exemplifies the great change which took place in the thinking of the nineteenth century. Agassiz was of the old school which thought that the world was a given which could be completely understood as it was. Darwin was an exponent of the newer school which felt that the world was dynamic and that in order to understand the world one must be able to explain its dynamic quality, which in itself causes one's understanding to be always subject to revision. Agassiz took nature passively; nature was majestic, to be enjoyed, and was rational and orderly. Darwin took nature actively; nature is powerful, to be conquered, and is irrational yet orderly, although this order must be found and constructed by man.

Agassiz was the last of a dying era; Darwin was a founder of the modern era. "Darwin's great fortune was that he was just enough ahead of his time to be a leader and not enough ahead to be ignored. Agassiz' misfortune was to have absorbed in his youth a *Zeitgeist* that was unsuitable for mixing with the revolutionary ideas."(62)

NOTES

1. William G. Land, "President Hill of Harvard on Agassiz and Darwin," *New England Quarterly*, VI (1933), 801. This article is primarily a reprint of Thomas Hill's "Louis Agassiz" which appeared in the *Unitarian Review.* December, 1885.

2. William James as quoted in Ralph Barton Perry, *The Thought and Character of William James* (Boston: Little, Brown & Co., 1935). I, 208-209.

3. Richard Hofstader, *Social Darwinism in American Thought* (Boston: Beacon Press, 1955), 11.

4. Max H. Fisch, "Evolution in American Philosophy," *Philosophical Review*, LVI (1947), 358.

5. In 1858, along with a letter to Asa Gray, an abstract of Darwin's

hypothesis appeared in the *Journal of the Proceedings of the Linnean Society: Zoology*, III (1858), 45f. On Darwin's own account this publication "excited very little attention." Francis Darwin (ed), *The Autobiography of Charles Darwin and Selected Letters* (New York: Dover Publications, Inc., 1958), 43-44.

6. Gavin de Beer, "Darwin's 'Origin' Today," *Natural History: The Journal of the American Museum of Natural History*, LXXV (Aug.-Sept., 1966), 62-71. This article traces the scientific developments which have proven the theory of evolution and which would be included in a new edition of the *Origin*.

7. For a general study of the events in science during this period see Bert James Loewenberg, "The Reaction of American Scientists to Darwinism," *American Historical Review*, XXXVIII (1933), 687-701, and Sidney Ratner, "Evolution and the Rise of the Scientific Spirit in America," *Philosophy of Science*, III (1936), 104-122. The former has an excellent bibliography, whereas the latter is more accurate in content, yet lacks bibliography. For events in Agassiz' life see Edward Lurie, *Louis Agassiz: A Life in Science* (Chicago: University of Chicago Press, 1960), 212-251.

8. See appendix which gives a chronology of these events.

9. Asa Gray, "The Origin of Species by Means of Natural Selection," *American Journal of Science and Arts*, 2nd Series, XXX (March, 1860); "Darwin and the Origin of Species," *Atlantic Monthly*, VI (July, 1860), (Aug., 1860).

10. Asa Gray, *Darwiniana* (New York: D. Appleton and Co., 1876).

11. Bert James Loewenberg, "The Reaction of American Scientists to Darwinism," 697. It is more accurate to say Gray's support overcame Agassiz' opposition, for Gray challenged Agassiz.

12. A. Hunter Dupree, *Asa Gray, 1810-1888* (Cambridge: The Belknap Press of Harvard University Press, 1959), 149, 247.

13. Edward H. Madden, *Chauncey Wright and the Foundations of Pragmatism* (Seattle: The University of Washington Press, 1963), 9.

14. These articles are inaccurately reprinted in Chauncey Wright, *Philosophical Discussions* (Charles Eliot Norton [ed], New York: Henry Holt and Co., 1878). Charles Norton was the editor of the *North American Review* at the time that these articles appeared; he considered Wright one of the outstanding philosophical minds in the United States.

15. For an excellent review of the religious literature of this period on Darwinism, see Bert James Loewenberg, "The Controversy over Evolution in New England," *New England Quarterly*, VIII (June, 1935), 232-257. This article contains an excellent bibliography. For additional discussion and bibliography, see Herbert W. Schneider, *A History of American Philosophy* (New York: Columbia University Press, 1946), 424-427.

16. Louis Agassiz, "Prof. Agassiz on the Origin of Species," *American Journal of Science and Arts*, 2nd Series, XXX (July, 1860), 142-154; "Evolution and Permanence of Type," *Atlantic Monthly*, XXXIII (Jan., 1874), reprinted in *The Intelligence of Louis Agassiz* (Guy Davenport [ed], Boston: Beacon Press, 1963), 215-232.

17. Francis Bowen, "Darwin on the Origin of Species," *North American Review*, XC (1860), 474-506.

18. John A. Lowell, "Darwin on the Origin of Species," *Christian Examiner*, LXVIII (May, 1860), 449-464.

19. Edward Lurie, *Louis Agassiz: A Life in Science*, 130-132.

20. A. Hunter Dupree, *Asa Gray, 1810-1888*, 286-287. It should be noted that Lowell's nephew, James Russell Lowell was the editor of the *Atlantic Monthly* at the time Gray's article appeared.

21. Asa Gray, "The Origin of Species by Means of Natural Selection," *Darwiniana*, 9-11.

22. Louis Agassiz, "Essay on Classification," reprinted, in its major sections, in *The Intelligence of Louis Agassiz*, 36-37, 109n3.

23. William G. Land, "President Hill of Harvard on Agassiz and Darwin," 797.

24. Louis Agassiz, "Essay on Classification," 38-39, 109n3, 109n4, 110n5.

25. This must be qualified by saying that Asa Gray in "Darwin and his Reviewers," *Atlantic Monthly*, VI (Oct., 1860), briefly discussed Agassiz, but missed the force of his position.

26. See appendix.

27. Edward Lurie, *Louis Agassiz: A Life in Science*, 58-59, 83; William G. Land, "President Hill of Harvard on Agassiz and Darwin," 796.

28. Francis Darwin (ed.), *The Autobiography of Charles Darwin*, 19.

29. William G. Land, "President Hill of Harvard on Agassiz and Darwin," 799.

30. Louis Agassiz, "Facsimile Letter of Professor Agassiz on the Darwinian Theory," *Scientific American*, XXX (1874), 87. This letter was written May 7, 1873 to Mr. James H. Parsons answering a question he raised concerning one of Agassiz' lectures entitled "The Egg."

31. Louis Agassiz, "Evolution and Permanence of Type," 219-220.

32. Louis Agassiz, "Prof. Agassiz on the Origin of Species," 154.

33. Chauncey Wright, "The Genesis of Species," (1871) reprinted in part in *The Philosophical Writings of Chauncey Wright* (Edward H. Madden [ed], New York: The Liberal Arts Press, 1958), 37.

34. Louis Agassiz, "Prof. Agassiz on the Origin of Species," 154.

35. Ibid., 145.

36. Charles Darwin, *The Origin of Species* and *The Descent of Man* (Two volumes bound in one, New York: The Modern Library, n.d.), 372.

37. Louis Agassiz, "Evolution and Permanence of Type," 230-231.

38. Charles Darwin, *The Origin of Species*, 372.

39. Ibid., 123.

40. Ibid., 64.

41. Louis Agassiz, "Evolution and Permanence of Type," 227.

42. Ibid., 222.

43. Louis Agassiz, "Prof. Agassiz on the Origin of Species," 147-148.

44. Agassiz acknowledges that a stock may be improved through selective breeding, but for him this phenomenon is the perfection of the species within the limits of that species; it is not the production of a new variety or species. See "Essay on classification," Section XV and 113n1.

45. Asa Gray, "The Origin of Species by Means of Natural Selection," *Darwiniana*, 25-30.

46. A. Hunter Dupree, *Asa Gray, 1810-1888*, 274.

47. Louis Agassiz, *Methods of Study in Natural History* (Boston: Fields, Osgood, & Co., 1863), 134. There is validity to this point, and it might be argued that Darwin did fall into this error. However, the proof of this point would be difficult to construct and goes beyond the purpose of this discussion.

48. Louis Agassiz, "Prof. Agassiz on the Origin of Species," 142.

49. Ibid., 143.

50. W. C. Wilson, "Darwin on the Origin of Species," *Methodist Quarterly Review*, XLIII (Oct., 1861), 613.

51. Charles Darwin, *The Origin of Species*, 43, 323.

52. Ibid., 323.

53. Louis Agassiz, *Methods of Study in Natural History*, 31.

54. Chauncey Wright was able to circumvent this difficulty; see below.

55. Louis Agassiz, "Prof. Agassiz on the Origin of Species," 143.

56. Ibid., 149, italics in text.

57. Ibid., 150.

58. Edwards H. Madden, *Chauncey Wright* (New York: Washington Square Press, 1964, 71.

59. This discussion is in Chauncey Wright, "Evolution by Natural Selection," *North American Review*, CXV (July, 1872), 1-30.

60. Wright had a better grasp upon the kind of explanation required to substantiate evolution than Darwin, but he unfortunately did not pursue the

line of thought he proposed as the correct answer. It should be also noted that Darwin was so pleased with Wright's analysis of this problem that he had it published as a pamphlet in England. See Edward H. Madden, *Chauncey Wright*, 72.

61. Asa Gray, "Design versus Necessity," *Darwiniana*, 70.

62. Ernst Mayr, "Agassiz, Darwin, and Evolution," *Harvard Library Bulletin*, XIII (Spring, 1959), 193.

APPENDIX

Agassiz' Participation in the Evolution Controversy: A Chronology

1845, June. Louis Agassiz writes a letter to Adam Sedgwick in which he states: "The idea of a procreation of a new species by preceding ones is a gratuitous supposition opposed to all sound physiological notions. And yet it is true that taken as a whole, there is a gradation in the organized beings of successive geological formations, and that the end aim of this development is the appearance of man. But this serial connection of all successive creatures is not material; taken singly these groups of species show no relation through intermediate forms genetically derived one from the other. The connection between them becomes evident only when they are considered as a whole emanating from a creative power, the author of them all."(1)

1857. Volumes I and II of Agassiz' *Contributions to the Natural History of the United States* are published. Volume I contains Agassiz' "Essay on Classification," which was reprinted as a separate volume in 1859.(2)

1858, December 10. Cambridge Scientific Club meeting. Asa Gray makes use of Darwin's theory in a paper on the distribution of Eastern North American-Eastern Asian plants.(3)

1858, December 14. American Academy of Arts and Sciences —Monthly meeting. Agassiz addresses the academy on the subject of the Classification of Fishes.(4)

1859, January 11. American Academy of Arts and Sciences — Monthly meeting. "Professor Gray gave a series of illustrations of the Botany of Japan in its relations to that of Central and Northern Asia, Europe and North America." This paper also makes use of Darwinian ideas. Agassiz and Gray enter into a discussion of the distribution of species. Agassiz also calls for special meetings "for scientific discussion" for February, March, and April.(5)

1859, February 22. American Academy of Arts and Sciences — Supplementary Meeting. Agassiz and Gray discuss the distribution of species.(6)

1859, March 10. Cambridge Scientific Club meeting. Agassiz speaks on "The Plan of Creation."(7)

1859, March 22. American Academy of Arts and Sciences — Supplementary Meeting. Dr. Charles Pickering reads a paper on the geographical distribution of species. Gray and Agassiz enter into a discussion.(8)

1859, April. The Museum of Comparative Zoology at Harvard is founded, under the directorship of Agassiz.

1859, May 12. Cambridge Scientific Club meeting. Asa Gray presents Darwin's theory of evolution to the club.(9)

1859, June. Agassiz makes a trip to Europe to buy specimens for the Museum of Comparative Zoology, thereby ending discussion of evolution for the year. Agassiz has returned by October.

1860, January 25. American Academy of Arts and Sciences — Stated Meeting. "Professor Agassiz, followed by Professor Parson, W. B. Rogers, and others [Gray is present] discussed several points in natural history and geology in reference to their bearing on the origin and distribution of species."(10)

1860, February 15. Boston Society of Natural History meeting. "Prof. Agassiz made verbal communication in opposition to the theory of Mr. Darwin, recently put forth in his work on the origin of species." Prof. Rogers answers Agassiz.(11)

1860, February 23. Cambridge Scientific Club meeting. Agassiz "reiterated his views on 'the number of distinct periods in the geology of the earth.' Since Agassiz emphasized the complete break between each stratum, Gray's paper to the club two weeks later on the axiom *Natura non agit saltatem* may be considered an answer and a flanking attack."(12)

1860, March 7. Boston Society of Natural History meeting. "Professor Agassiz made a communication on consecutive faunae and their corresponding number of geological formations, as furnishing arguments against the theory of Darwin." Professor Rogers answers.(13)

1860, March 21. Boston Society of Natural History meeting. Professor Agassiz continues his discussion on faunae of March 7th.(14)

1860, March 27. American Academy of Arts and Sciences — Special Meeting. "Mr. J. A. Lowell, Professor Bowen and Professor Agassiz discussed adversely the hypotheses of the origin of species through variation and natural selection. The latter reiterated his denial of the existence of varieties, properly so called, in the animal kingdom, in the wild state, insisting that what were so called were either stages of growth or phases of cyclical development, and the domesticated varieties were of no account in the matter, having no counterpart in nature."(15)

1860, April 4. Boston Society of Natural History meeting. Professors Agassiz and Rogers discuss Darwinism.(16)

1860, April 10. American Academy of Arts and Sciences — Monthly Meeting. "Professor Gray criticized in detail several of the positions taken at the preceding meeting by Mr. Lowell, Professor Bowen, and Professor Agassiz...."(17)

1860, May 1. American Academy of Arts and Sciences — Special Meeting. "Professor Gray resumed the discussion upon the questions mooted at the preceding meeting....The discussion was continued by Professor Agassiz and others."(18)

1860, July. The *American Journal of Science and Arts* publishes "Prof. Agassiz on the Origin of Species," which is from advance sheets of Volume III of Agassiz' *Contributions to the Natural History of the United States.*(19)

1860, October 9. American Academy of Arts and Sciences — Monthly Meeting. "Professor Agassiz reiterated his opinion that what are called varieties by naturalists do not in reality exist as such." Professor Gray answers.(20)

1862. Agassiz completes and publishes Volume IV of *Contributions to the Natural History of the United States,* which is the last he finishes.

1863. Agassiz publishes *Methods of Study in Natural History,* which he considers a complement to his "Essay on Classification."(21)

1865. National Academy of Science. During the year Professor Agassiz read "On the Individuality among Animals, with reference to the Questions of Varieties and Species," "On the Metamorphoses of Fishes," and "On the Geographical Distribution of Fishes, as bearing upon their Affinities and Systematic Classification."(22)

1865, April 4-21. On shipboard to Brazil for a collecting expedition, Agassiz delivers a series of lectures to his assistants in which he soundly goes against evolution.(23)

1867, October 16. Boston Society of Natural History meeting. In response to an invitation by the president of the society, Jeffries Wyman, Agassiz offered some remarks on the antiquity of man. The *Proceedings* does not indicate if any discussion followed.(24)

1872, October 24. At San Francisco, Agassiz gave a lecture on his Hassler Expedition which was reported in the British journal *Nature*. On Agassiz' discussion of Darwin, they state, "We need not follow the speaker through his singular misrepresentation of the theory of evolution."(25)

1873, July and August. Just after the opening of the Anderson School of Natural History under the directorship of Agassiz, he gave a series of lectures in which he again discusses Darwin. He closes one lecture entitled "Basis of Classification" with: "Do we recognize in the manifestations of physical forces, as heat and electricity, anything indicative of thought and combination? Is there anything that shows that the lowest [organisms] should be first? Do we not have in this series which I have indicated an evidence of intelligent action which we do not find in physical forces? Nothing but a presiding intelligence could produce these wonderful gradations from the simple to the complex. Therefore I am opposed to Darwinism."(26)

1873, December 14. Louis Agassiz dies.

1874, January. *The Atlantic Monthly* publishes Agassiz' "Evolution and Permanence of Type."(27)

NOTES FOR APPENDIX

1. Elizabeth Carey Agassiz, *Louis Agassiz: His Life and Letters* (Boston: Houghton, Mifflin and Co., 1886), I, 392.
2. A basic chronology of Agassiz' life is found in Guy Davenport, *The Intelligence of Louis Agassiz*, 233-235.
3. A. Hunter Dupree, *Asa Gray, 1810-1888*, 252.
4. Proceedings, IV, 108.
5. Proceedings, IV, 133-134, 136.
6. Proceedings, IV, 171.
7. A. Hunter Dupree, *Asa Gray, 1810-1888*, 258.
8. Proceedings, IV, 184.
9. A. Hunter Dupree, *Asa Gray, 1810-1888*, 258.
10. Proceedings, IV, 362. This is the first mention of the *origin* of species.
11. Proceedings, VII, 231-232.
12. A. Hunter Dupree, *Asa Gray, 1810-1888*, 285.
13. Proceedings, VII, 241.
14. Proceedings, VIII, 250.
15. Proceedings, IV, 410.
16. Proceedings, VII, 274-275.
17. Proceedings, IV, 411.
18. Proceedings, IV, 424-426.
19. Louis Agassiz, "Prof. Agassiz on the Origin of Species," *American Journal of Science and Arts*, 2nd Series, XXX (July, 1860), 142-154.
20. Proceedings, V, 72.
21. Louis Agassiz, *Methods of Study in Natural History* (Boston: Fields, Osgood & Co., 1863).
22. *Annual of the National Academy of Science for 1865*, 39-40. A lack of funds prevented these papers from being published.
23. Professor and Mrs. Louis Agassiz, *A Journey to Brazil* (Boston: James R. Osgood and Company, 1867), 3-45.
24. Proceedings, XI, 304.
25. Anonymous, "Agassiz at San Francisco," *Nature*, VI (Oct. 24, 1872), 509-510.

26. All these lectures were reported by the *New York Tribune*, and then republished in book form with other science articles by the *Tribune* as *Popular Science* (Boston: Henry L. Shepard & Co., 1874); quotation at 53.

27. Louis Agassiz, "Evolution and Permanence of Type," *Atlantic Monthly*, XXXIII (Jan., 1874), reprinted in *The Intelligence of Louis Agassiz*, 215-233.

Western Illinois University,
Macomb, Illinois.

The Fall and the Field

by James F. Sheridan

One of the continuing themes in Sartre's *Critique of Dialectical Reason* is that dialectical reason can and apparently must constrict itself to one of its "moments", that is, to analytical reason. This fall may be simply a shift from reasoning in a way which is not wholly precise to reasoning in a way which restricts itself to the methods of mathematics, logic and some forms of empirical inquiry. That sort of reasoning is an instance of what Sartre calls *la Raison analytique.* But the account is incomplete because the first "step" in an analytical process is not to break down a complex into its elements nor to reason upon atomic elements but is instead to bring into being a realm, e.g., a *champ pratique*, which is amenable to such treatment. Actions of this sort are neither mysterious nor far to seek. Measurement of a physical phenomenon is a device to enable us to treat it mathematically. To "consider" the bodies of our solar system as unextended points is an instance. The establishment of translation rules which permit us to associate the symbol for extensional implication with combinations like "if, then" as used in the natural language is a third example. There is no need to multiply cases. Human beings certainly find it useful to treat portions of their world so as to make those portions available for precise investigation and manipulation, indeed, they sometimes find it necessary to do so —for certain purposes. That they do not always do so is a deliverance of common experience. If we are to believe what the analysts have discovered, then speaking "ordinary English" is a massive instance of a not-wholly determinate form of life. But the point here is that one *can* convert to greater linguistic precision. In these respects and only in these respects, there is an analogy between the relation of ordinary English to a technical mode of speech and the relation between dialectical and analytical reason. The move from the less definite to the more definite arena can be and is made in both contexts. But one major difference between them must be cited immediately. It was said of the transition from one variety of reason to the other that it can and "apparently must" occur where the only sense of 'must' exhibited was "necessary —for some purpose". Nothing was or should have been said about this characteristic with respect to the natural language. It is the intent of this essay to try to puzzle out what is meant by 'must' when predicated of the "fall" and what this implies about the "field" of which dialectic is to be the "reason".

Let us begin by reflecting upon a circumstance apart from the context of systematic inquiry in which we frequently "prepare the field" so that it admits of definite action, namely, a situation in which we are faced with a decision which is or which appears to be crucial. To minimize the possibility that we merely concoct a case to fit a preconceived description, let us consider the end of an old love affair, since most adults will have been through that experience and will be well enough beyond it to remember with indulgence and perhaps with fondness. If we consider one that occurred when we were sufficiently young, little need be said about the poignancy of the experience. The reader's memory will provide what is required. Reflection suggests that there was a time in all of our lives when our world was ordered around a young woman both in the sense that our other activities were ranked below our life with her in degree of importance and perhaps even in the sense that we depended upon her to make up for whatever lacked in the rest of our existence. The terror which we can now see as an origin of the fury of the break suggests our degree of commitment. The making and unmaking of the decision to part; the exhilaration of considering alternatives coupled with the sudden self-consciousness brought by finding oneself in fact or in prospect "on the market"; the lengthy and frequent conferences with well meaning friends who tried to patch up the quarrel; all these and more recall the comparative chaos into which our lives were cast. One remembers the relief felt which was often expressed as, "Well, at least the decision is made. Now I can get back to a normal existence."

The expression in quotations is suggestive. A decision had in fact been taken. But the language also suggests that there remained a lingering suspicion that it was not entirely well founded, that there were tag ends, that there was an element of the arbitrary clinging to the selection of the solution which somehow was unwanted in a decision of such moment. The unease felt then can be emphasized if one remembers how inconclusive were the discussions in which one tried alone or with others to come to a "rational" decision. While many discussions, taken individually, were ended with a definite view, there were almost always more of them, a factor which suggests that the definiteness was very localized. We were frequently able to "make sense" but the whole process did not make sense. Further, the very occurrence of continued reflection and conversation emphasizes the feeling that the elements considered did not "add up" to the decision and that an effort at rationalization was required. "It was all for the best" is a remark that can be made only after much time passes or after an alternative center of organization has been found.

What this tale is intended to suggest is that there are moments in the experience of most when for a period of time, *le monde vécu* has the tint of madness. One hopes that most men have had several moments of this kind. But one cannot be so sanguine if one turns to other instances of other types of insanity, ranging from decisions of a similarly crucial kind in other aspects of life to the experiences of those who have been through deep therapy. No rehearsal of these is required. What is in point there is that the frequency of such moments is much greater than one ordinarily admits to himself or to another. If a practicing therapist read that line, he would howl at "admits". "Permits" would be preferable. This tone is also suggested by one of the quotations above: "Now I can get back to a *normal* existence". Normal? Perhaps. Most will assert that the greater part of their daily lives lack the kind of encounter with contingency which is discussed here. Everyday life is surely more like the period before the love affair than it is like the time of the death throes. Relative order is surely more *usual* than relative disorder. But to say this is to come perilously close to a merely statistical claim, a mere concept of "average". It is all too easy to slip from the relief that periods of relative stability are more frequent and of longer duration than periods of relative instability to the belief that they constitute normality and perhaps even sanity or reality. Clearly such periods do not constitute *totality* since they are identifiable only because their contraries do in fact occur. One may wonder, then, why periods of instability should be regarded as lapses, as regrettable, as anomalies, unless they are taken only in a statistical sense. For instance, since *one* of the familiar senses of 'real' is "vivid" or "intense", and since we regard emotions as frequently intense and since emotional intensity is frequently associated with periods of instability, one wonders why 'real' is not asserted of periods of disorder. There is no intent here to argue this case. Rather, these considerations are introduced to pave the way for entertaining the supposition that the domain of relative disorder is prior to that of relative definiteness both in the order of knowledge and in the order of being.

Let us so suppose. What one would assert is that we "begin" in a realm which is not wholly definite, a realm much more like that which phenomenologists call the "pre-objective" than like the world of "normalcy". Perhaps we might even say that if moments of crisis are taken seriously, we would discover that the everyday world as we live it is precisely a *response* to a more fundamental level of apprehension whose traces remain in the everyday world much as the traces of the troublesome elements of the death throes of a love affair remain in the world constructed through the decision to end the affair. In such a constructed world, Sartre can speak of an ego which is "in the world" rather than "in" consciousness. But "constructed" is a poor term. Among other things, it is an idealist's term. 'Posit' or 'reveal' would be more Sartrean. But he would also say that the unity by virtue of which we call this a "world" is *synthetic*, a unity of not wholly similar entities. Better yet, the world in which one finds himself and that self constitute entries in a mixture of disparate elements. If one finds this intolerable, what is done is to make the situation more definite. Through *emphasizing* one or more of the disparate elements, one suppresses some immediate exhibition of contrarity. Relative stability is achieved. Permanence is the result of maintaining some product of some such emphasis whether that permanence be attained through mere persistence or through deliberate commitment and continued effort. Any subsequent situation will have such products as ingredients but there is apparently an inexhaustible reservoir of intermixed contraries. There are those elements which have been suppressed, for instance, and there is the opposition between the product and its less organized "origin". Making-definite, it seems, does not avoid contrariety but simply accommodates it in different forms than it "originally" took.

But it must be added immediately that one is seldom faced with sheer flux. Rather, one begins from some previous product, some *other* form of organization. That is why I keep putting quotes around the term, original. The relative flux, the other organization, exhibits not only the surpassed form but a coefficient of adversity. There are relations among the entities which are independent of any particular *depassement.* The products of my *praxis* have consequences which I did not envisage. Sartre remarks that *praxis* by definition includes ignorance and error as among its fundamental structures. But to say this is to say that one learns in and through praxis that there are relationships *synchroniques* which are not wholly malleable. They are revealed only in and through *praxis* but they are *être-pour-dévoilé*. Do I ever deal with pure possibility? No more than I ever deal with pure flux. The first move has always been made when I enter the game. If my *usual* life-world is that of some form of relative stability, if, as suggested here, the action of the free being which I am is action influenced by the various *situations* in which I find myself, if I encounter resistance —as I must if I am to speak meaningfully of freedom or of *praxis*— then nothing would be more natural than to suppose that the usual is the *given*. From the point of view we are now considering, however, that cannot be. From this stance, it would be much more plausible to emphasize Sartre's endorsement of the Marxist contention that man is the product of his product. Thus one would contend that having brought a world into being, man becomes the product of this product and begins to live the field through the constituted world. As long as man maintains that style of life, he will select, arrange and act from within it even though he may always abandon it.

To emphasize the influence of a style of life is surely one of the major reasons that Sartre wrote *CRD*. It appears that he accomplished his goal so well that many of his sympathetic critics have accused him of having abandoned the "existentialism" of *EN*. That sort of extreme judgment indicates that those critics had not fully understood the strictures upon freedom which were presented in *EN*. There, as here, the human reality was always *in* a situation, a situation which included elements which were *not* chosen and included dimensions and consequences which were not foreseen. Sartre's point has always been that the index of human freedom was man's ability to take account of that with which he was presented without being forced to consider himself as a function of the presentation. Where freedom was directed to the content of the presentation, Sartre has tried to show that some of the elements were there only because some individual put them there. But notice —*some*. Where human freedom was to the situation considered as a whole, it has been Sartre's contention that only a being who was ahead of himself, only a being with the permanent possibility of transcendence *could speak* of the "whole", or, in the language of *CRD*, totalities occur only within totalizations. But despite this sort of critical reminder, one must grant to the critics that *EN* left one with the lingering suspicion that only God could have written the book as only God could make the claim that Marxists make when they assert the rationality of the real. In *CRD*, this is not a problem. It is clearly an historical being who wrote this book. Man is now exhibited as that being who earlier was only *said* to be "always already in the world". It is as though having "overstated" human freedom in *EN*, Sartre is now paying the dialectical price and has moved to the opposite "extreme". Only the grossness of such architectonic restrains us from pressing the point.

The supposition which we are considering is that the non-definite is to be understood as prior to the definite but if *CRD* is as massive a statement of immersion in what is relatively definite as the preceding paragraph indicates, the chief difficulty would seem to be how we could ever surpass the definite to glimpse beyond. This essay has suggested that moments of "crucial" decision afford such glimpses. The heavy volume of literature devoted to the presentation of "encounters" provides a catalogue of more colorful occasions for such grasp. Sartre did not speak of his "literature of *extreme* situations" without deliberate intent. But we also know from his earlier stance that the fact of an encounter does not entail any particular response nor any particular *kind* of response. One may seek to go on authentically or one may fall back into the absurd. Sartre told us as early as 1945 that repression was the most usual response, that we live most of our lives in a bad faith. In *CRD*, we have a lengthy statement of the difficulties one experiences in seeking to avoid continued bad faith, in particular, the difficulties encountered because man is not only a free being but a *social* being. The last phrase is doubly suggestive. Sartre's commentators have detailed the extent to which man's sociality constitutes a barrier to his freedom. They have done less to emphasize the perennial Sartrean claim that resistance is a necessary condition for the realization of freedom. But that is not the point here. It is the fact that man is *both* free *and* social which interests us. If the "weight" of man's sociality is now presented as greater than has been the case in Sartre's earlier works, that does not eliminate his freedom nor make his struggle to give it maximum expression any less intense.

Sartre has used many formulas to call attention to this doublet. "Man is what he is not and is not what he is." "Man is in the world and conscious of it." "The human reality is always being-in-a-situation." Freedom and resistance are correlative terms. The doctrine of intentionality which binds consciousness and its "object" together inseparably carries the force of the contention. "L'homme, c'est une passion inutile." The passion was useless because it was the desire to be the complete coincidence of consciousness and the object of consciousness. Does it require the standpoint of deity to make such assertions? The central effort of *CRD* is to show that it does not. Sartre's intent is put by him in a Kantian manner so that no mistake will be made: *Assuming that there is one history, how is it possible for a being who is himself historical to know this?* Merleau tells us that the Hegelian view of totality could only have been taken by deity. Perhaps Sartre, too, learned this from Hegel as he certainly learned much else. For instance, he learned that Spirit actualized itself in modes which were so different from itself that it often failed to recognize itself in them. The roots of both Sartre's and Marx's versions of alienation lie here. But neither man could countenance a notion of Spirit in itself either as a virtuality or as the end of history nor could Sartre admit that individuals possessed no autonomy other than that which was required for the inevitable progress of Spirit. From within history, man was found to be ambiguous, paradoxical, absurd, and, in *CRD*, a being caught in the web of his own creation and *therefore* responsible for its maintenance or its surpassing. Whether or not he would surpass any "world" is a matter in which sheer contingency is ingredient. But the field itself he could not surpass because it was the condition of there being worlds to be surpassed. To use another formula —being is contingent— and necessarily so.

But to say this is to begin to argue that neither complete definiteness or complete indefiniteness is necessary or even possible. What is necessary is their conjunction. I shall call this the "non-definite" or the "field". It is this which is apprehended in moments of crisis either after the shock of experiencing the dissolution of the definite has worn off or perhaps across a series of such shocks. It is this to which the famous descriptions in *Nausea* were directed. It is no accident that Sartre chose to devote so much of his early efforts to literature, for what better mode of presentation than the literary could be chosen for the invitation to the non-definite? One sees how much of his technical philosophical production has been rejected by Anglo-Saxons —and 'rejection' is used deliberately here— because the attempt to treat the non-definite discursively has resulted in paradox and epigram and a concern with generality which is very much out of vogue west of the English channel. It is also no accident that the bulk of Sartre's efforts in the last ten years have been devoted to non-literary production. Even the book on Genet is a philosophical treatise. What is suggested here is that Sartre has gone beyond the stage of the *experience* of the non-definite, beyond the stage in which it is useful to use literature to evoke that sort of awareness. The major effort must now be to deal with the non-definite by "intellectual" means but to deal with it in its own terms. To say what this means, let us suppose that the proper means of apprehension of the indefinite is "insight" and the proper mode of apprehension of the definite is analytic reason. Then one might say that the kind of experiential grasp exhibited in the literature was a contemporary version of the mystical apprehension of the field. Through this grasp, the invitation was extended to experience the field as indefinite or to experience the indefinite aspects of the field while keeping some of its definite structures in focus serially. Then one might regard *EN* as an attempt to grasp the field analytically. One can make outlines of *EN*. Desan did. One can draw spatial diagrams as one can with traditional systematic philosophical works. The book is shot through with paradox which is exactly what one would expect if the categories of a definite mode of treatment were applied to a not-wholly definite domain. If the field was revealed directly through the mystic grasp, it is revealed indirectly through the shipwreck of analytical reason in *EN*. No claim is made here that Sartre's first major work *must* be read in this fashion but only that it can be read thus with profit for some purposes.

In the same spirit, one is tempted to suggest that the literary productions, the essays, like "Materialism and Revolution", and *EN* were Sartrean "reductions" in the sense that they made the field available to scrutiny by a variety of special interests. Professional eggheads could enter through *EN* and because it was replete with "descriptions" as well as being thoroughly argued, could proceed to consider and perhaps to accept the invitations offered through the literature. The entrance through literature offered so many access to the field that one need only mention that move. It is well if the essays are added to lead those who entered through the literary portal to a more substantial grasp of content but it is necessary to confine oneself to an *invitation* when one deals with the devotees of experience for they frequently have no desire to do anything other than experience. The "progression" of course is not necessary. One can stop with the literary works and bemoan the fact that so little has occurred of late. One can also stop and write scholarly works on *EN*. But one could also find that studies of these works could lead to the attempt to deal with the non-definite not from the kind of stance appropriate to one of its elements but through a mode of apprehension which resulted from the interiorization of other modes. The motive for such a move would be that one had gone through both of the other modes and found the reason for going beyond them in the very process of developing them. One might even argue that one would have to proceed in this manner for the consciousness of the dialectic must itself be dialectical. The motive for transcendence of any given stage can only be found by immersing oneself in that stage. It could scarcely be otherwise if an historical being is to speak of one history. To stand apart from the modes of apprehension themselves and give an account of them would be to take the standpoint of deity with respect to knowledge and that stance is no more permissible in epistemological matters than it is when the theme is that which is apprehended.

We do not intend to suggest that one could not deal competently with *CRD* unless he had read the literary works and *EN* although a strong case could be made for that view if it were Sartre's brand of dialectic which was in question. That there are other avenues of approach than his is patent. But it is claimed that the apprehension of a non-definite domain must itself be non-determinate, that the realm of dialectic can only be "known" dialectically. If one adopts the Sartrean view that the object apprehended and the mode of apprehension are not wholly distinguishable from each other, one asserts the existence of a *field* in which to develop one is to develop the other. The totalization of a detotalized totality is itself creative, that is, constituting and constituted in the sense that what is surpassed is the product of a previous surpassing and that the resultant totality is not wholly reducible to its elements. Totalities make sense only within a *totalisation en cours*, causal relations are revealed only with the compass of a means-end rubric, analysis makes sense only given a prior synthesis. When the field in which one operates is shot through with contraries which range from the basic opposition of the definite and the indefinite (*en-soi* and *pour-soi* ?) to the fact of a multiplicity of totalizers, what is required is a "method" which can encompass such diversity without recourse to a level of generality which would be featureless, a method which, as Merleau said, could integrate unreason into reason. For Sartre this injunction has come to mean the action of a material being in and through the environment which he and others of his kind have formed and do form, in and through a world made on the basis of antecedent conditions but not reducible to those antecedents, in short, through the totalizing action which man is.

But is there such a field? And if there is, how do we account for the fact that some men become "arrested" in the definite and some become lost in the indefinite? Let us consider the question as to the existence of the field first. Those who emphasize the indefinite aspect would have us believe that in moments of crisis, the field is directly apprehended. We are invited to participate directly or vicariously in the experience of which this is the deliverance. Note that the claim is not made that any such experience would constitute *evidence* for the existence of the field any more than one can properly say that direct perception constitutes evidence for what is perceived. Neither does any question arise which would suppose that the frequency of the experience or the scope of the agreement of the presumed experiencers is germane to the issue of existence. Like mystical apprehension, this pre-mystical grasp is its own certification. Here as in the case of dialectic, "necessity" is necessity *for* some individual but is not wholly peculiar to him. To say that something can only be certified by the experience of an individual is not to say that what is grasped is *possessed* by him and only by him. What all this chatter is about is an attempt to account for experiences which are said to be self-certifying, the claim to which is most often made by "prophetic" existentialists. But if the experiences are self-certifying, one ought to stop justifying and start experiencing —or failing to experience as the case may be. When the partisans of the indefinite tell me that something is the case and that I certify it if I experience it and when I don't experience it, I'm unenlightened, I begin to suspect chicanery. But apart from this familiar problem there is another and more serious difficulty. If mysticism is the logical extension of experiencing appropriate to the indefinite, then to proceed in this fashion is to embark upon a quest which if done thoroughly would yield a domain which is featureless or which at least lacks the kind of features which allow us to *speak* of it. Thus the emphasis upon the experiential apprehension of the field would not yield the field unless it were restrained short of its natural fulfillment. But such restraint would have to come from something other than the experiencing itself so that only "enlightened" insight would be the ground of assertions concerning the existence of the field. One ought, then, to look for the source of enlightenment since experience alone is not enough.

The obvious option is to consider whether an emphasis upon the definite would allow the assertion of the existence of the field. One thing can be said immediately for this approach. It grants the existence of the indefinite and thus the conjunction of what are said to be the two fundamental structures of the field. To clarify is to grant that there is the unclarified whether that be the object of inquiry or the domain which includes the definite object and the confused grasp which was had prior to clarification. Consider, for instance, the claim that a considerable advance was made in the understanding of the field when animism was abandoned and we began to treat our world as a puzzle, that is, to carve out a domain to be called "nature". To say that one treats it as a puzzle is to say that one deals with it as though it were "over there", to treat it as though one did not interact with it or in it, in short, to minimize the importance of subjectivity. This would seem analogous to Sartre's account of the transformation of an ambiguous domain into a partial *champ pratique* through making oneself to be an object in that domain. If this is also true, that one *can* apply the puzzle approach to humans and their relations but that one *need* not (whether one *should* aside), then the field is loose enough not to be intrinsically a puzzle although structured enough to permit one to treat many, most, or all parts as puzzle elements. In fact, to compare the modern treatment of Nature to animism is precisely to grant that the field is sufficiently non-definite to admit of either approach. If it is granted that some have treated the field as something other than a puzzle, then however one characterizes that treatment, whether as illusion, error, or superstition, one grants that the field may be thus treated even though there are penalties attached. Thus one grants that the field does not have *only* those characteristics which permit us to regard it as a puzzle. If "there is" both appearance and reality, the field is not *only* what is called "real" but also includes appearance whether that be an ontological structure or the result of the finitude of the humans to whom "things" appear.

One may conclude that to emphasize the definite is to assert that the field exists, for the presupposition of the definite is the indefinite and the conjunction of the two is the field. But "conjunction" seems too weak a term for what has been claimed here as well as too weak a term for the claim that the calculus appropriate to the field, dialectic, intermittently constricts itself to one of its moments, the analytic. After all there are people who stay indefinitely in the indefinite pose, there are those who are committed to logical analysis, and there are dialecticians. Why should an ingredient of the field, a human, restrict himself to what is but a moment of the most fundamental mode of living it? To the question, how is it possible that such a transformation should occur, Sartre has a ready answer. Analytical reason *is* a moment of dialectical reason. Thus there is no problem *in principle.* But how and why it occurs *in fact* is a problem of the same stature for Sartre that the famous "application problem" is for Kant. Let us consider some of Sartre's offerings. He cites practicality. Since the field has partial orders, one deals best with them in their own terms. To use dialectic in empirical inquiry is at least inefficient. Second, Sartre has suggested that direct confrontation with nothingness is unbearable, so difficult that to live with contingency present constantly to mind, to "act without hope", to achieve authenticity, is the most difficult human accomplishment, always precious, always transient, never finally achieved. This claim cannot be *argued,* of course. One can only issue the invitation to experience —or to remember vividly. Encounters are tropical growths. But Sartre's chief contribution to this issue is the weight which he has assigned in *CRD* to the difficulty of achieving full freedom in a social world conditioned by scarcity. One cannot read the first volume without acquiring a sense of inevitable doom, of fruitless, pointless struggle whose chief virtue is that it can be taken up again and again. The viscous quality which Sartre ascribed to his world in *Nausea* haunts the pages of *CRD.* To make oneself to be an object in order to obtain purchase in a world appears to add to the "mass" of the organism so that one thinks of *aging* as the image most appropriate to Sartre's descriptions. For instance, that I am a body is not due to me. That I am a six foot, four inch body is only indirectly my doing. But that I am a *fat* six foot, four inch body is largely due to me. But it is also true that all I need do to maintain the lard is to relax. My body is now in a condition where it will fatten of its own accord if I do not actively oppose it. It is as though it had a tendency independent of my will, or, better, a tendency which is something in itself which can be affected or even arrested by my effort but which will continue if I don't intervene. One can notice recognize Sartre's notion of *sedimentation* in this familiar and non-special context. If one added similar commentary about one's psychic tendencies, one might well say that the self he is becoming becomes increasingly dense, that the processes by which one is affected and which one wishes to affect become increasingly imperturbable. Since one is "always already in the world" by the time one begins to reflect, one has already generated a rather "substantial" self and that self continues in part because of reflection and in part on its own. The object which one has always been continues to be and although reflection may be ingredient in the "later" selves to a greater extent than in the earlier, the later selves are even more substantial than the earlier. A precise illustration would be to say that my possibilities at 39 are more limited than they were at 29. They are limited *because* of the increasingly heavy thing I become, the corresponding increase in difficulty in moving it, the diminution of energy due to the extra effort required just to stay even, and the disappearance of some real possibilities both as to quantity and as to accessibility.

What has been said immediately above indicates that one makes himself to be an object or —better— emphasizes his object-being in order to obtain purchase in a field of objects and that the object he becomes acquires an unexpected degree of autonomy. I committed myself to becoming a college professor fifteen years ago and the object which I have become grows steadily more formidable. Despite my relatively casual ways, I have increasing difficulty in distinguishing Jim Sheridan from Professor Sheridan. Sartre goes much further in his analysis than this since he tells us that our materializations return against us — the anti-dialectic ensues. Thus part of the answer to the question "How is it that one actually becomes an object? ", is that once one has begun the process, the process acquires a life of its own and the increments become greater without further deliberate intervention. In Sartre's language, *praxis individuelle* becomes *praxis-processus.* "It" is maintained only in the sense that it is not actively opposed as one maintains anti-Semitism merely by failing to actively oppose it. But one must add to this partial answer that one "originally" emphasized one's object-being as a matter of practicality. One can even envisage a situation in which unusual foresight was available so that one could grant the claim that if he committed himself to an otherwise unpalatable activity, he would increasingly take on the characteristics common to those who thus immerse themselves. It might still be said that the price was worth it. Thus one might chain himself to a typewriter day after day in order to surpass his current grasp of a field even though those he trusted made it all too clear that to escape from those chains would in time come to be as close to an impossibility as one can claim without being in hilariously bad faith.

But both of these partial answers are unsatisfactory because both suggest a degree of contingency to the process of becoming an object which is appropriate only to becoming a certain *kind* of an object. Sartre's general structures no more entail a particular response than they have ever done despite the appearance of such specificity given to *CRD* by Sartre's commitment to Marxism, or, as I should prefer to say, to the "left liberal" style of life. The very form of the question, "How does the transformation from the not wholly definite (dialectical reason) to the almost wholly definite (analytical reason) actually occur? ", begs the question. It suggests that there may be a "time" when we are not actually in a process of totalization so that one might choose not to objectify himself. But if, as Sartre has it, the human organism *in a field of scarcity* is totalizing, if the organism is the activity itself as including its products, one cannot ask how it began to do so. It is for this reason that the term "emphasize" was used in talking about the commitment to definiteness. But to talk this way is to be more faithful to Sartre than Sartre is to himself. He tells us that an organism by its very nature is a harmonious process of self-fulfillment, a cyclical renewal, but because it occurs in a field of scarcity, that rhythm is interrupted and need and *praxis* are thus structures of an organism in a field of scarcity.

The claim that organisms objectify because they find themselves in a domain characterized by scarcity is most provocative. One applauds Sartre's consistency in making necessity a consequence of contingency since the necessity of objectification is a function of the contingent occurrence of scarcity. But one cannot help but hesitate about several aspects of this contention. Although Sartre can hold his claim in the face of the fact that some organisms fail in their attempt to regain equilibrium —some, in fact, must fail— it is clear that some *human* organisms *choose* to refuse further objectification. There are suicides. That fact does not counter Sartre's assertion that all organisms objectify themselves in so far as they are organisms but it does raise again the question of how such an entity can arrest itself. This question appears to be the permissible form of the question about how an individual can begin to objectify. One fears that Sartre might do something atrociously familiar like appeal to an "instinct" for self-preservation or characterize suicides as "abnormal" and thus dismiss them as candidates for counter-instances. Even if one supposes that it is a "fact" that most men struggle to survive, that struggle acquires an aspect of contingency beyond that which it possesses by virtue of its situation just because there are those who abandon the struggle. This reference to additional dimensions of contingency raises the second hesitation. As is well known, Sartre tells us that our world is conditioned by scarcity but he adds that the world need not be and he verbalizes the possibility of conceiving a world of plenty. It has been suggested elsewhere that scarcity may well be the guise under which the contingency of being is exhibited in the world we happen to have. It has also been said that scarcity is not only a lack of material goods although it is illustrated as such in the greater part of *CRD.* Nor is it surprising that in a work on man as a social being that the readiest illustrations should be drawn from the realm of material "necessities". It is quite likely that a Marxist of any variety would regard such factors as fundamental to the domain of human activity and perhaps a good part of what is meant by Sartre's claim that Marxism is the philosophy of our time is precisely that it accords this fundamental status to "economic" factors. This observation is consistent with Sartre's assertion that the revolution is permanent, that one can speak of a time when the world will be such that Marxism will no longer be its philosophy. But one has to ask how strictly Sartre wishes to hold to *material* scarcity, for if the philosophy of our time is currently in a *cul de sac*, the original innocence is lost and one can no longer give Marxist answers to questions about Marxism. One must ask, for instance, whether a revolution which would overthrow the current social, economic and political structures of our world would be enough to eliminate scarcity. Sartre has not abandoned his eternal vow of hatred against the bourgeoisie nor his views concerning the reasons for the importance of the struggle for the underdeveloped nations. There are many indications that he might well regard the *socialist* revolution as the means for the elimination of scarcity. This "left liberal" who has made common cause with the Communist Party because it was the most serious and the most effective of the action sections of the left may well have difficulty in distinguishing his commitment to man from the immediate means of effecting that commitment. But we prefer to suppose that the notion of scarcity is more fundamental than the notion of lack of material goods just as we prefer to suppose that Sartre's commitment to man is more fundamental than his commitment to the Party. It is Sartre who has said that the revolution is permanent, that man the totalizer cannot become man the totalized. If this is so, scarcity will not be eliminated by radically altering a given set of social-political arrangements but only a certain mode of scarcity will be surpassed. Perhaps we will be able to eliminate arrangements which systematically deprive humans of the opportunity to achieve the necessities of existence without becoming sub-humans and few who think this possible would refrain from seeking to actualize it in their own fashion. But to say that one can imagine a world from which scarcity *per se* was absent would be almost to say that one can imagine a world from which contingency was absent and *that* Sartre cannot say. I suggest that Sartre so fears the characteristic liberal commitment to principles which are *merely* ideal that he is overly reluctant to distinguish scarcity in general from scarcity in its current forms. This motivation would not be new to Sartre since it was a large part of the motivation which led to the intractable tract, *The Communists and the Peace*, the essay which so infuriated Merleau-Ponty and many other less acute members of the "non-communist left". No doubt some of them would still object that Sartre's failure to distinguish the general from the particular is quite independent of his particular political commitments and instead is merely an instantiation of his view that the general is seen only through some particular and thus cannot be distinguished as sharply as these comments seem to require. To this one can say that Sartre has never held back from general characterizations of man and although they were no doubt *presented* —and perhaps developed— through prior descriptions of particulars, they were asserted apart from their sources. It could scarcely be otherwise for a man who refuses all forms of reductionism. Whether one "derives" the claim, "man is a useless passion" from particularities or invents particular cases to illustrate the general, the general is not wholly reducible to the set of particulars. No more is asked here with respect to the notion of scarcity.

What is desired of Sartre is that he clearly connect the notion of scarcity with what one wishes to argue are more fundamental structures of his position, where that position includes *EN* and *CRD* as members of a single framework. The rapidity of the transition from the discussions of organisms to that of individual *praxis* in *CRD* suggests that Sartre is not unaware of this need at least to the extent that a *human* organism will be both in the world and conscious of it and thus will continue to exist as a lack which partially reflects the necessary contingency of Being. It may well be that one must take individuals as "mediations", as particularized universals, as *singulars* so that in *anthropology*, one will tend to emphasize what is peculiar to the individual or even to groupings or regions of more restricted generality than those which were the themes of the phenomenological ontology called *EN*. One is tempted to say that *CRD* is a "regional ontology" in Husserl's sense of the term. But scarcity will be on the line between any such regional effort and general ontology and Sartre makes it very difficult to see that this is so. Part of the reason for that obscurity is that *CRD* is a joint result of Sartre's ontological labors and the form which his commitment to man takes, that is, to Marxism. No doubt he would prefer to say that Marxism is the combination of ontology and existentialist "descriptions" but since it is Sartre's variety of Marxism which is in question here and not conventional Marxism, one suspects that such an objection would be of little continuing consequence apart from the commitment to man. On the other hand, one could make a case for the possibility of understanding *CRD* as a dialectical development from *EN*, the "abstract" first work, but to make that case through anything but the mere imposition of an architectonic would require precisely the kind of explicit generalization of the notion of scarcity which this discussion faults Sartre for failing to provide. Perhaps the pending second volume will do just that although the announcements concerning its contents give little reason to expect it. One wishes Sartre would write a third major work.

There are two other hesitations concerning these issues which bear commentary. The first may be introduced by recalling what has been said thus far about the descent to definiteness. When it is said that one makes himself to be an object because he finds himself in circumstances which require that response, one is nagged by the feeling that he is faced with an explanation of the form, "Sleeping pills put one to sleep because they have a soporific power", or, "Things are what they are —specifically— because they are what they are —generally". Not only is it *possible* for men to make themselves to be objects because they have an object-being but they actually do so and, indeed, must do so because part of what they are is beings in a field of scarcity. One sniffs an "argument" of the form, "Men are totalizers. Totalization includes objectification as a moment. Men thus necessarily objectify themselves." What went wrong? No claim is made that this is a kind of structural problem which is peculiar to Sartre for, if nowhere else, it is apparent in the work of his master, Hegel. One remembers that it was Hegel who spoke thus in insisting that the system was its own warrant, that the certification of propositions in the system was available only to those who would work through it. The same could be said of Sartre's brand of dialectic. If we take seriously the claim that the second volume of *CRD* is to present us with the maximum opportunity to test the dialectically grounded claims, one may have to pay the same exhaustive and exhausting price for that examination which one must pay if one takes the prolific Hegel seriously. But the first volume will permit some commentary in Sartre's defense that is independent of the requirement of immersion in his writings which some will resist. Sartre tells us that he tries to speak in *CRD* of one history from the standpoint of an historical being. One meaning of that claim is that the various structures and components discussed will be the result of an analysis of the synthesizing in which we find ourselves. As was the case with Kant, part of what Sartre intends by calling his efforts "Critique" is that the moments are displayed as the results of analysis. If this approach is adopted, the suspicions cited above may be understood to arise from the mistake of taking the products of analysis as the foundation of synthesis. One remembers that the modern period in the history of philosophy gives us several examples of men who took subject and object as the fundamental framework within which all synthesis was to take place thus failing to realize that these two notions were products of the analysis of processes in which they were merely moments. That period also taught us —or at least it taught both Kant and Hegel— that to proceed in this fashion was to be unacquainted with one's own intentions and led to insuperable problems which need not even have occurred apart from the faulty methodology, much as paradoxes like that of the raven need not arise unless we insist on interpreting the copula in terms of extensional implication. If critique leads Sartre to hold that we can account for the totalities and structures we encounter in "experience critique" —including objectification— only by positing totalization, one must assess that contention *as critique* and not by treating it as though it were a "question of fact". But Sartre must write the second volume as Kant should have written the first edition of the *Critique* before he had written what was in fact the second edition.

We turn now to the second hesitation. To say that to be a totalization is to constitute totalities and that this process of constitution is a necessary condition of *dépassement* is faithful to Sartre's stance but again suspicious. The suspicion is heightened if one notes that it is entirely plausible to understand Sartre to hold that man finds himself in a world in which his objectifications acquire an autonomy of their own, an autonomy which has the force of resisting the process of surpassing. The dialectic results in anti-dialectic where the very products of free action restrict that freedom. Freedom further objectifies itself —and by now it should be clear that we must always say "further"— and becomes subject to ossification because resistance is freedom's correlative and this champion of radical freedom must also be a champion of radical resistance. If the resistance encountered were only that which is the reflection of the real in the world, freedom would have to be less radical than Sartre took it to be in the period of the plays and of *EN*, the period whose central theme some say has been lost in the "radical conversion" which is *CRD*. A freedom so "intense" must have a product which *actively* opposes it as well as the passive resistance which is the mark of the *en-soi.* There must be an anti-dialectic. But that counter force must also be the *product* of freedom if freedom is to be given its due. A freedom so intense must be the source of its very degradation so that, so to speak, it may seek itself beyond its most powerful foe, itself as embodied. It is true that in *CRD*, we are presented with reams of material which exhibit the resistance provided to the individual because he is a social being, because he is one among a plurality of totalizers. It is even tempting to attribute to this neo-idealist Royce's view that if resistance is the criterion of the real, the human, if you like, the social, is of greater dignity than the merely material. But for Sartre, it must be the individual himself who is his own worst enemy in the very context in which he is most his own. This theme bears considerable expansion but paragraphs can be saved by a single comparison. It is not Orestes of *The Flies* who is the Sartrean hero —it is Hoerderer in *Dirty Hands.* Freedoms can become arrested *because* men must be thus if they are to be fully free. It is only in the surpassing of *opposites* that the dialectic leads to full freedom. The human reality *still* is not what it is and is what it is not.

Now it is time to ask why so much space has been devoted to the field and to action when this essay purports to be a discussion of the "fall" of dialectical reason to analytical reason. To begin to answer, let us remember that the first step in analytical reasoning is to establish a domain in which it may be used. That establishment is an act within a totalization and it is in this sense that analytical reason is a moment of dialectical reason. But this is to say that neither analytical nor dialectical reason is an element of a "methodology" in the sense in which one sometimes distinguishes deductive and inductive reasoning as methodological constituents. One could make the mistake of considering this analogy apt only if one *neglected* the "first step". One could *teach* analytical reason within the compass of this neglect. If one then took analytical reasoning as the model and regarded dialectical reason as a different procedure of a similar kind, then one might ask about the domain in which it was to be properly used, the techniques through which it could be applied, and the like. There are courses in deductive and in inductive logic. *But there are no courses in dialectic.* Dialectic, as Johnstone clearly shows, is essentially a bilateral activity and is learned only in confrontation whether that confrontation be direct or vicarious. To represent Sartre correctly, however, we must emphasize "activity" and understand that to be predicated of a material being. Neither of the two kinds of reasoning are done apart from the life-world, neither are done from a neutral standpoint or a spectator view. To speak of reasoning is to speak of the life-world which it clarifies and modifies. To speak of reasoning is to speak of the *field* in which it is ingredient.

But to speak of the field is not quite the same as to speak of a realm in which analytical reason is efficacious, for there is no claim that dialectical reason brings the field into being. Dialectic is the "logic" of the non-definite. But it might have been said with equal justice that dialectic is the order of the field. Because that might just as well have been said, the question as to whether dialectic is a procedure or an ontological structure can arise, the debate over whether or not there is a dialectic in Nature can ensue. Sartre denies that Nature is dialectical except in the sense that the positing thereof is part of the process. But he insists that history is dialectical and it is clear that one can read back to *EN* and say that the "worlds" we constitute are not different in this respect. But the field is inclusive of worlds and analytical domains and the various sub-structures of these arrangements. The field is non-definite because it includes these definite groupings and it includes the indefinite, the margin which is the ground of the possibility of other ordering or further ordering. It is the field, the non-definite, which we have always with us. When we "clarify" it, we frequently say that the result of the clarification is how it "really" is. That's false. If one must talk in this fashion, what endures is the field including the various arrangings humans contribute to it. The field seen through but one of those arrangements is "abstract" in the Hegelian sense.

Because dialectic is *both* a logic and an order, because it is both constituting and constituted, because the very awkwardness of the sentence, "Dialectic is the reason of the field" points up the commingling, one may be tempted to treat dialectic as a nisus independent of the humans who exemplify and use it. It was this aspect of the dialectic which Merleau emphasized and it is because of that emphasis that he flirted constantly with the sort of "fetishizing" to which orthodox Marxists have fallen prey. Sartre took this clue seriously enough to provide us with the prospect of totalizations without totalizers, seriously enough to make the explication of collectives a central task in *CRD.* But he refused to allow himself to be drawn far enough from his earlier "voluntarism" to sacrifice the human individual to an ontological idol. History makes men *precisely* in so far as men make history. The price which Sartre pays for insisting upon this fundamentality of the human individual is to speak often in paradox and thus to verge upon betraying his own insight into the character of the field by using the language of the partisans of the indefinite. He also tempts one to ask the question which runs through these pages, that of how it is that the dialectic may be arrested, by giving the process a degree of autonomy which makes one dubious about the possibility of effective individual interference. No claim has been made here that either Sartre or this essay satisfactorily accounts for this puzzle although it is true to say that Sartre's chief weapon against the autonomy of the process is the autonomy of the individual —and vice versa. Perhaps it is best to leave the issue where it is with but one further remark. The claim is made that this puzzle is a product of Sartre's own proceeding and we have tried to show how this is so. But is it possible that in order to see this issue as a puzzle one requires in addition to Sartre's proceedings a commitment to analytical reason? Is the "satisfactory account" referred to above a surreptitious request for a degree of definiteness which is inconsistent with the nature of the issue? It is not clear that this is so but since it may be so, let us by all means leave the issue here.

Allegheny College,
Meadville, Pennsylvania.

Aristotelian Influences and References in Harvey's *De motu locali animalium*

by Jerry Stannard

An important milestone in the history of science and medicine was passed in 1628 with the publication of *De motu cordis* by William Harvey (1578-1657).(1) By this single, small book of seventy-two pages, the circulation of the blood was irrefutably established upon experimental evidence. Since that time, Harvey has been regarded as one of the earliest advocates of the experimental method.(2) He has, for the same reason, been honored in text books as "The Father of Modern Physiology." An unceasing flow of articles and books devoted to one or another aspect of Harvey's accomplishments has brought him undiminished praise, and occasionally, uncritical adulation.(3)

It is certainly not my intention to cast doubt upon Harvey's well-deserved place among the luminaries nor to call into question the significance of his discovery. The superlatives lavished on *De motu cordis*, however, have tended to obscure other features of Harvey's earlier work and his methods. As Pagel has recently noted, "less attention has been paid...to the development of the discovery [*scil.* of the circulation of the blood] in Harvey's, the discoverer's own mind."(4)

In many respects Harvey was a typical medical practitioner of the late Renaissance. He was, after all, an Elizabethan physician, sharing the beliefs and presuppositions of his age. Following an early education at King's School, Canterbury (1588-1893) and at Gonville and Caius College, Cambridge (1593-1599) he enrolled at the University of Padua in 1600.(5) The most famous medical school in Europe at the time, Padua excelled in the teaching of anatomy based on the dissection of corpses.(6) In this, she carried on a tradition going back to Vesalius who did for anatomy what Harvey was to do for physiology. But Padua also had been the seat of Aristotelianism, professed by such teachers as Pietro Pomponazzi (1462-1525), Jacopo Zabarella (1533-1589), and Cesare Cremonini (1550-1631).(7) At Padua, Harvey studied under Hieronymus Fabricius ab Aquapendente (1533-*ca.* 1619) the renowned anatomist and an undeniable stimulus to Harvey's search for the true function of the valves in the veins.(8) Fabricius for all his debt to Vesalius and his successors, was also a fervent Aristotelian and Galenist and unwilling to adjust his philosophic commitments to the "New Science."(9)

Harvey received his diploma in 1602 and returned to England where he became a Fellow of the Royal College of Physicians in 1607. Shortly thereafter, he was appointed physician to St. Bartholomew's Hospital in London and entered into a career of teaching and medical practice, numbering Francis Bacon and Thomas Bodley among his many patients.(10) Upon the death of Thomas Davies, Harvey assumed the additional duties of the Lumleian Lecturer and began delivery, in 1616, of a cycle of lectures. Later, referring to these lectures, he stated in the dedicatory epistle of *De motu cordis* that

On several earlier occasions in my anatomical lectures I revealed my new concept of the heart's movement and function and of the blood's passage round the body. Having now, however, for more than nine years confirmed it ... by numerous ocular demonstrations, and having freed it from the objections of learned and skilful anatomists, I have yielded to the repeated desire of all and the pressing request of some, and in this small book have published it for all to see.(11)

Harvey's own notes for these lectures, the *Prelectiones Anatomiae Universalis*, still survive and have recently been published.(12) The passage relating to his discovery of the circulation of the blood, translated from his curious mixture of Latin and English, runs as follows:

WH From the structure of the heart it is clear that blood is constantly carried through the lungs into the aorta *as by two clacks of a water bellows to rayse water.* By [application of] a bandage [to the arm] it is clear that there is a transit of the blood from arteries into veins, wherefore the beat of the heart produces a perpetual circular motion of the blood *(unde perpetuum sanguinis motum in circulo fieri pulsu cordis)*(13)

The words immediately following the above passage are not without their own interest. They indicate that at the time of writing — sometime between April 1616 when he became Lumleian Lecturer and January 1618 when he carried out his second year's anatomical dissection — Harvey had not freed himself from the Aristotelian-Galenic conception of the heart and its function. He asked:

Is this for the purpose of nourishment, or more for the conservation of the blood and the parts by the distribution of heat; and, in turn, is the blood cooled by warming the parts made warm [again] by the heart? (14)

Little need be added here concerning Harvey's later career beyond noting that in 1649 he replied to the most able of his critics, Jean Riolan the Younger (1580-1657)(15) and, well along in years, he published his last book, the *De generatione animalium*, in 1651.(16)

Throughout Harvey's published writings, references to his own experiments and observations, often preceded by the monogram WH, alternate with references to the writings of his predecessors.(17) By and large, he is harder on his more immediate predecessors than on the ancients. His opposition in *De motu cordis* to Aristotle and Galen has been widely publicized as sounding the death knell for the outmoded methods and erroneous theories then dominating European medicine.(17*a*) There is no doubt that the Aristotelian-Galenic theory of circulation was dealt a mortal blow by Harvey. After 1628 it was increasingly difficult to maintain the many mutually inconsistent portions of the Aristotelian-Galenic theories of circulation and respiration.(18) Through a superb combination of experimental and quantitative evidence, the older doctrine of *spiritus* was effectively eliminated from cardiovascular physiology.(19)

At some undetermined time but clearly later than 1616 at which time Harvey began to write up his Lumleian Lectures, he was compiling another set of notes for a projected book on the movement of animals. These notes were not completed or rewritten in final literary form. In the words of his editor, "they remain the rough draft for Harvey's intended book."(20) The whereabouts of these notes is unknown until the time they entered the British Museum as part of the collection formed by Sir Hans Sloane (1660-1753). Presumably they had been lost along with Harvey's notes and manuscripts when his apartment at Whitehall Palace was looted in 1642.(21) Twice in the middle of the last century, the existence of an unpublished Harvey manuscript was noted, first by Willis(22) and somewhat later by Paget.(23) But even Power's study at the close of the century did not dispel uncertainty regarding the provenance and date of this manuscript.(24) Because MS Sloane 486 contains two sets of notes, one on the muscles, the other on animal movement, it was assumed that the date 1627, occurring on folio 70 beneath the heading *De motu locali animalium*, referred to the manuscript as a whole. That this is not the case has recently been demonstrated by Whitteridge who has shown that the date 1627 refers only to the second part of the Sloane MS.(25)

The year 1627 represents, then, the year in which Harvey began to assemble notes on animal movement. It is, however, difficult to determine when he stopped adding notes or revising them. Two passages in *De motu locali*, though important as indices of Harvey's method of composition, unfortunately fail to establish a precise date. Twice Harvey refers to a *Librum de motu cordis* which can only be the title of his book published in 1628.(26)

Several possibilities exist for explaining and dating these two references. On the one hand, they may have been written in 1627 when Harvey, by his own admission, was firmly in possession of the principles of the circulation of the blood. That was a time when conceivably he was applying the finishing touches to *De motu cordis* in time for it to appear for the Frankfort Book Fair in 1628.(27) On the other hand, the two passages in question may have been inserted later, *after* the publication of *De motu cordis.* (28) But whether they were written before 1628 or afterwards, they indicate something of the importance of the notes for purposes of gaining a better understanding of Harvey's development.

The significance of *De motu locali* is enhanced by the fact that it is a draft of a study in which the millennium-old doctrine of *spiritus* bears much of the burden. We are faced, that is, with notes which demonstrate that Harvey, on the eve of publishing his epoch-making book *or after* was not prepared to reject Aristotle.

There is nothing surprising in the fact that Harvey would rely so heavily on Aristotle. His published writings as well as the unpublished *Prelectiones* testify to Aristotle's influence. It could, of course, be argued that the *Prelectiones* represent a stage in Harvey's development that was abandoned in *De motu cordis* and *De circulatione.*(29) But to maintain this position is to minimize the tentative date of *De motu locali* and to overlook the Aristotelianism of Harvey's *De generatione animalium.* While Aristotle is cited dozens of times in *De motu cordis*, sometimes approvingly, those passages probably represent a literary convention rather than a genuine debt to Aristotle's methods or data. At the very least it seems safe to say that the authority of Aristotle did not blind Harvey to the significance and full exploitation of his experimental evidence.

In *De motu locali*, however, the references to Aristotle are of a different sort; certainly they do not represent a literary convention which would be inappropriate in a work book. Moreover, experimental data are scanty in *De motu locali* and there is no indication that Harvey had made any quantitative studies on animal movement parallel to the famous argument of *De motu cordis*, chapter IX. These notes, rather, look back to a time when science and medicine were little more than scholia to Aristotle, written within the framework provided by Aristotle's philosophy. In the present case, Harvey is indebted to Aristotle for the very formulation of his problem. The notes begin abruptly with these words:

> Motus species: alteratio, quali;
> augmentatio, quanto; localis, ubi.(30)

That this is virtually an abbreviated paraphrase will be seen by comparing this passage with Aristotle.(31)

From Aristotle, Harvey also borrowed some of his terminology and illustrations, possibly some of his evidence concerning the movement of different species of animals, and derived general support for his conclusions by noting their agreement with Aristotle's position. Like his Aristotelian-oriented predecessors, Harvey did not restrict his reading to the biological writings. In addition to *De motu animalium*, *De incessu animalium*, *De partibus animalium*, and *Historia animalium*, Harvey also cites *Physica*, *Problemata* and the spurious *De spiritu*.

Aristotle is cited by name forty-five times in *De motu locali*. This figure, however, gives no indication of the extent of Harvey's debt. For there are, in addition, numerous other passages which are close translations or paraphrases of the Latin edition of Aristotle but lacking an explicit reference to his name or to the title of one of his writings. Of Aristotle's biological writings, Harvey most often refers to *De motu animalium* and *De incessu animalium*. It may appear strange that such emphasis is placed on two of the smallest works in the *Corpus Aristotelicum* until it is recalled how well their text accorded with Harvey's viewpoint and the state of his research.

The important point to note is that Harvey's approach is determined by Aristotle's formulation of the problem. Although other authors are cited, notably Galen, Fabricius and even his contemporary Robert Fludd, *De motu locali* must be judged as an index of Harvey's thinking, not as a published book which in any way changed the direction of science or medicine.(32)

The problem of determining how animals move is largely a theoretical exercise in Aristotle's *De motu animalium*. In the preceding *De partibus animalium*, he collected the factual data and arranged them in order to demonstrate the causes exhibited by each part. Moreover, in the *Physica*, to which Aristotle himself refers, it had been established that all types of movement or change derive ultimately from an unmoved mover.(33) Thus, in

De motu animalium, Aristotle can turn from the ultimate causes of movement to a consideration of the causes of, or conditions for, local movement of animals. As a methodological note, he remarks that questions concerning movement must be confirmed by reference to individual organisms. The starting point of such an investigation, he continues —and Harvey repeats this very advice— must be sense perception.(34)

This rather abstract introduction to the study of the movements of animals is tacitly adopted by Harvey but justified in a manner which finds no direct parallel in his sources. Local motion is necessary, he argues, for purposes of food-getting, reproduction, and preservation or survival.(35) Local motion of the parts, for example, phototropic responses of plants and the movement produced by the passage of *spiritus* in breathing and in the pulse, presupposes that the organism as a whole is capable of undergoing local motion.(36) This is accomplished by tonic movement *(motus tonicus).*

The elementary physiological unit thus turns out to be tonic movement and about it Harvey has much to say, though largely derived from Aristotle and Galen. Although Aristotle does not seem to use the word *τονικός* (37) the basic idea is similar to the claim that overt movement is effected by a change of *spiritus* or *pneuma* (*πνεῦμα*). In *De motu animalium*, the effects of *pneuma* in the soul are likened to the function of a joint of a limb.(38) Both require an immovable, substantial basis in order to be effectual as the means by which or from which the motive power is transmitted. In the case of the joint of a limb, which both Aristotle and Harvey recognize to be an application of a fulcrum, this is a stationary support, for example, the ground when walking or running, the air for the flight of birds or water for the swimming of fishes, etc.(39) A further example drawn from Aristotle and repeated by Harvey concerns the ease by which a boat is moved by a pole provided that the thrust originates from outside the boat "but if he tries to do this when he is inside the boat itself, Tityus could not move it nor Boreas by blowing from inside it."(40) In the case of the soul, the original motive force is a combination of the physical effects of *pneuma* and a projected end or goal of action. How Harvey understood this will be discussed below.

Relying, then, on the adequacy of the Aristotelian theory of *pneuma*, Harvey attempted to translate this into the anatomical and physiological terminology of his day. Unfortunately for Harvey, the conceptual framework of early seventeenth century anatomy and physiology had made little progress since the time of Aristotle and Galen. Hence Harvey had little recourse than to try to adapt the older ways of thought to his growing collection of empirical data.

This was accomplished primarily by tracing the consequences which followed from the basic assumption that muscular contraction is a form of tonic movement.(41) There are two stages by which Harvey adapts the older theory of tonic movement to his particular needs. First, the capacity of a muscle to contract or to exhibit any form of tension is due to the effects of *spiritus* and secondarily of heat.(42) But *spiritus* and heat (or vital heat) are only the material cause; they must be transmitted to the requisite muscle or tendon, the effective cause, before overt action can occur.(43) This leads Harvey to take a step which, whether he was hesitant or not, seems difficult to justify on anatomical grounds alone. That is, he is led to assume that a nerve, the conductor of *spiritus*, is like a pipe in its construction; it does not itself move, but merely supplies the material source of tonic movement.(44) Although Harvey seems to vacillate between sound anatomical evidence, that the nerve itself does not move except *per accidens*, and the erroneous claim, that the nerve is a hollow conduit, he makes use of his clinical training to justify his position. In a query following the first of the two passages cited immediately above, he notes that when a nerve be damaged, a spasm ensues.(45) It would thus appear that Harvey, the experimental physiologist and comparative anatomist, is seeking empirical data to fit the Aristotelian scheme he has adopted. There is a precedent for this, of course, for it is well known that Aristotle's contributions to biology were sometimes made at the expense of adjusting his empirical evidence in order to fit a preconceived philosophic scheme.(46)

Harvey's method may also be illustrated by his pneumatic interpretation of the various organs and their respective functions. The muscles, as was noted above, are capable of contraction or tonic movement because of the *spiritus* conveyed to them by the nerves.(47) Presumably the blood also helps to transport the *spiritus* throughout the body for it is specifically stated that *spiritus* is set in motion by the heart and flows through the arteries.(48) In another passage equally striking for one who can lay claim to having called into question the adequacy of a pneumatic explanation of circulation, he states that the purpose of the veins and arteries is to nourish the body.(49) Whatever this may mean, it is further evidence of Harvey's method, at once both deductive and inductive, Aristotelian and experimental. As we know today, it was not until after Harvey's death that Malpighi (1628-1694) succeeded in demonstrating by the use of microscope that the blood's transit is completed by the capillary anastomosis between arteries and the veins.(50) Harvey guessed at the existence of the capillaries and the logic of *De motu cordis* required that the circulatory cycle be continuous. But in 1627 Harvey seemed to be a long way from realizing the momentous consequences which would follow from his discovery. Otherwise it is difficult to imagine why he would claim that the purpose of the veins and arteries was to provide nourishment or the similar statement that the beating of the heart was due to the *spiritus*.(51)

The innate *spiritus*, in addition to being the source of movement, played an important role in generation, pathological anatomy and mental activity. Harvey does not spell out in detail his thoughts on these matters but sufficient hints appear in isolated passages to trace their Aristotelian ancestry. After having referred to *spiritus* as implanted, Harvey is ready to assign further functions to *spiritus* in accordance with his source.(52)

Since *spiritus* is the medium between body and soul,(53) it follows that just as the soul is ultimately responsible for movement of the whole organism, so the parts are moved by *spiritus*.(54) Birth and death, the two terminal stages of life are likewise explained in terms of the nature and function of *spiritus*. Semen contains *spiritus* while death apparently is the exhaustion of the implanted *spiritus*.(55) This is not explicitly stated in *De motu locali* but from the fact that *spiritus* is breathed forth(56) and that *rigor mortis* is the state characterized by lack of muscular contractility,(57) it appears that Harvey would follow his sources here as elsewhere. Another indication of his tendency to follow his sources occurs in his explanation of abnormal convulsions. These, he states, are due to *spiritus* acting contrary to nature.(58) Even the distinction between higher and lower forms of life seem to be restatable in pneumatic terms for Harvey once notes that some animals possess only *spiritus*, apparently the necessary condition of life.(59)

Returning now to mental activity which Harvey no less than Aristotle, considers as a function of the soul, it is maintained that *spiritus* is both the organ of movement and of feeling.(60) Desire, and apparently this would hold true for other forms of mental activity, is excited or put into motion by *spiritus*.(61) Muscles react or contract as the result of a message derived from the soul,(62) though elsewhere Harvey accords primacy to the brain. Writing of the latter as the *mester del choro*, or choir-master, he likens the function of the nerve to a process of adjudication.(63)

Numerous other allusions to and borrowings from Aristotle doubtless could be located by further study. But it must be remembered that *De motu locali* is only a series of rough notes for an intended book. How Harvey would have put his notes together into a coherent whole and what the finished product would have resembled, can only be guessed at. Nor can it be determined whether, as a result of re-examining his anatomical evidence and following to its logical conclusion the central argument of *De motu cordis*, he would have parted company with the doctrine of *spiritus*. As a matter of fact, Harvey did not take this decisive step until the letters to Riolan.(64)

For the purpose of relating his data at the preliminary stage, *Spiritus* was a useful hypothesis and one capable of explaining a wide range of biological processes and behavioral patterns. If the concept of *spiritus* turned out later to be an *ignus fatuus*, it was due, in large measure, to Harvey himself. But before this could come about Harvey, like all pioneers, had to struggle to find the correct direction.(65) *De motu locali*, then, is a precious document, illustrating Harvey's mind at work and, incidentally, one reason why the doctrine of *spiritus* lasted as long as it did.

NOTES

1. *Exercitatio Anatomica de Motu Cordis et Sanguinis in Animalibus.* Frankfort: William Fitzer, 1628. For the editions and translations, see Geoffrey Keynes, *Bibliography of the Writings of Dr. William Harvey, 1578-1657.* Second Edition. Cambridge, 1953.

2. "...the greatest of England's early experimental scientists". Geoffrey Keynes, *The Life of William Harvey.* Oxford, 1966, p. 1.

3. Robert Grove (1634-1696) Bishop of Chichester, is an example of an effusive and not-too-critical eulogizer. Cf. L.R.C. Agnew, De Sanguine Episcopoque. *Bulletin of the History of Medicine* 34(1960) 318-330.

4. Walter Pagel, An Harveyan Prelude to Harvey. *History of Science* 2(1963) 114-125 (ref. p. 114).

5. For biographical details I have followed Keynes (note 2, above).

6. Cf. Klaus König, Die Stellung der Anatomen unter den medizinischen Lehrern in Padua im 16. Jahrhundert. *Centaurus* 7(1960) 1-5.

7. Cf. John Herman Randall, Jr., *The School of Padua and the Emergence of Modern Science*, Padua, 1961, especially pp. 71-114.

8. Cf. K. J. Franklin, Valves in Veins: An historical survey. *Proceedings of the Royal Society of Medicine* (History of Medicine Section) 21(1927) 1-33.

9. Cf. *The Embryological Treatises of Hieronymus Fabricius of Aquapendente* with an Introduction, a Translation, and a Commentary by Howard B. Adelmann (Ithaca, 1942) pp. 82 sq. and the review of the above by Max H. Fisch, *Bulletin of the Medical Library Association* 31(1943) 376-379.

10. Cf. E. F. Mauer, Harvey in London. *Bulletin of the History of Medicine.* 33(1959) 21-36.

11. William Harvey, *The Circulation of the Blood and other Writings.* Translated by Kenneth J. Franklin. London, 1963. P. 5.

12. *The Anatomical Lectures of William Harvey. Prelectiones Anatomie Universalis: De Musculis.* Edited, with an Introduction, Translation, and Notes by Gweneth Whitteridge. Edinburgh and London, 1964.

13. William Harvey, *Lectures on the Whole of Anatomy.* An annotated translation of *Prelectiones Anatomiae Universalis* by C.D. O'Malley, F.N.L. Poynter, and K.F. Russell. Berkeley and Los Angeles, 1961. P. 191. I have followed the editors by retaining Harvey's English words and phrases in italics. For the monogram WH, see below, note 17.

14. *Lectures on the Whole of Anatomy* pp. 191-192. An expression of Harvey's debt to Aristotle is the occurrence of a paraphrase of a passage from *Historia animalium* I 16, 494b21-24 on the first folio of the *Prelectiones.* For a facsimile of this folio, cf. *Lectures on the Whole of Anatomy* p. 8.

15. *Exercitatio Anatomica De Circulatione Sanguinis.*a *Ad Joannem Riolanum Filium Parisiensem* .... Cambridge: Roger Daniel, 1649.

16. *Exercitationes de Generatione Animalium.* London: Typis DuGardianis, 1651. Cf. A.W. Meyer, *An Analysis of the De Generatione Animalium* of William Harvey. Stanford, 1936.

17. The monogram WH "appears to indicate that he [sc.Harvey] personally, as contrasted to others, has observed or can vouchsafe the particular phenomenon." *Lectures on the Whole of Anatomy.* p. 19. Another symbol, Δ, occurs below in note 45. This has been interpreted to mean something Harvey has demonstrated or plans to demonstrate.

17*a*. Cf. Erna Lesky, Harvey und Aristoteles. *Sudhoffs Archiv* 41(1957) 289-316, 349-378.

18. Cf. Donald Fleming, William Harvey and the Pulmonary Circulation. *Isis* 46(1955) 319-327 and Leonard G. Wilson, The Transformation of Ancient Concepts of Respiration in the Seventeenth Century. *Isis* 51(1960) 161-172.

19. Cf. F.R. Jevons, Harvey's Quantitative Method. *Bulletin of the History of Medicine.* 36(1962) 462-467.

20. William Harvey, *De motu locali animalium.* Edited by Gweneth Whitteridge. Cambridge, 1959. p. 3. Hereafter cited simply as *De motu locali* plus page reference. On occasion I have in the following notes supplied Harvey's own words in the belief that the reader will prefer to make his own translation and/or interpretation.

21. Cf. Keynes, *op. cit.* p. 162 (see above note 2) who cites a passage from

Harvey's *De Generatione animalium* describing his losses.

22. William Harvey, *The Works.* Translated from the Latin with a Life of the Author by Robert Willis. London, 1847.

23. George E. Paget, *Notice of an Unpublished Manuscript of Harvey.* London, 1850.

24. D'Arcy Power, *William Harvey.* London, 1897. p. 67.

25. *De motu locali* pp. 2-3.

26. Vide librum De motu cordis. *De motu locali* p. 20; Vide librum De cordis, an motus vitalis. *ed. cit.* p. 44.

27. Cf. E. Weil, William Fitzer, the Publisher of Harvey's De motu cordis, 1628. *The Library* 24 (1943-1944) 142-164.

28. Mrs. Whitteridge points out in the *apparatus criticus* that the reference to a *librum De cordis* on p. 44 was "written in a differently coloured ink with a different quill and in a somewhat tidier hand than most of the MS, [and] was evidently added on a subsequent occasion". But there is no indication that the similar reference on p. 20 was not made at the time of writing.

29. Cf. J.S. Wilkie, Harvey's Immediate Debt to Aristotle and to Galen. An Essay Review of William Harvey, *Prelectiones Anatomie Universalis* ... ed. G. Whitteridge. *History of Science* 4(1965) 103-124.

30. *De motu locali* p. 14.

31. *Physica* V 2, 226a23-25 especially 226a 24-25: *λείπεται κατὰ τὸ ποιὸν καὶ τὸ ποσὸν καὶ τὸ ποὺ κίνησιν εἶναι μόνον.*

32. Cf. A.G. Debus, Robert Fludd and the Circulation of the Blood. *Journal of the History of Medicine* 16 (1961) 374-393.

33. *De motu animalium* I. 698a 8-12; in this passage Aristotle refers back to *Physica* VIII 5, 25864-9.

34. *De motu animalium* I, 698a12-15. Cf. *De motu locali* p. 32.

35. *De motu locali* p. 16.

36. *De motu locali* p. 18.

37. The one occurrence of *τονικός* in Aristotle, *De partibus animalium* IV 12, 693b12 has recently been disputed by Peck who conjectures *πτητικός*. Aristotle, *The Parts of Animals* with an English Translation by A.C. Peck. London, 1955. p. 406n.

38. Cf. Aristotle, *De motu animalium* X 703a4, especially a12-13: *τοῦτο* [*sc. τὸ πνεῦμα σύμφυτον*] *δὲ πρὸς τὴν ἀρχὴν τὴν ψυχικὴν ἔοικεν ὁμοίως ἔχειν ὥσπερ τὸ ἐν ταῖς καμπαῖς σημεῖον, τὸ κινοῦν καὶ κινούμενον, πρὸς τὸ ἀκίνητον.*

39. *De motu locali* p. 26. Cf. Aristotle, *De motu animalium* II, 698b11-20.

40. Aristotle, *De motu animalium* II 698b24-26. Translated by E.S. Forster (in) Aristotle, *The Parts of Animals* (see above note 37). Cf. *De motu locali* p. 54 for Harvey's use of this illustration.

41. Utraque eadem actio musculi contractio et motus tensio sive motus tonicus. *De motu locali* p. 116.

42. Vires, intention, etc. a spiritu et calore. *De motu locali* p. 128.

43. Cf. Harvey's extended discussion, entitled *Moventium Differentiae* in which he adopts the Aristotelian causes. *De motu locali* pp. 38, 40. On p. 38 he writes: Corporalia, Spiritus, Calor Naturalis, instrumentum instrumentorum.

44. Nervus ut canalis nullo modo movetur. *De motu locali* p. 106.

45. Δ si [sc. nervus] vel contusus, vulneratus, refrigeratus, calefactus, [excitantur] spasmis, deliria, febris acuta etc. *De motu locali* p. 106.

46. Cf. Edwin Clarke and Jerry Stannard, Aristotle on the Anatomy of the Brain. *Journal of the History of Medicine* 18(1963) 130-148.

47. An per nervos ad musculos spiritus tanquam nutrimentum sensi. *De motu locali* p. 100. Cf. p. 108: WH sed an spiritus a cerebro per nervos, [vide] ante in 'Spiritu'.

48. Spiritum per arterias influere motum ex corde: 1 sanguis et spiritus una res, et musculus et spiritus motivus, unde nutriri spiritus ut corpus. *De motu locali* p. 102.

49. Venam et arteriam ad nutriendum quia molles partes facile difluant et opus nutrimento maiori. *De motu locali* p. 92.

50. Marcello Malpighi, *De pulmonibus observationes anatomicae.* Bologna: J.B. Ferroni, 1661.

51. A spiritu fit pulsus cordis. *De motu locali* p. 94.

52. The phrase *spiritus implantatus* occurs without comment in a list of "parts" recognized by Aristotle as pertaining to muscle. *De motu locali* p. 64. As Harvey does not appear to use the phrase again, it is not completely clear what he intended.

53. quia dilatari et contrahi potest [sc. motus spiritus] eque ad utrumque pulsum tractum, unde medium inter animam et corpus...*De motu lacali* p.94.

54. Sicut animal ab anima movetur ita corpus a spiritu ita musculus, ... *De motu locali* p. 96.

55. *De motu locali* p. 41. For Harvey's opinions regarding the causes of death, cf. *Lectures on the Whole of Anatomy* pp. 170, 174, 176.

56. ... spiritus ex parte oris. *De motu locali* p. 140.

57. Musculus mortuus non contractile, stiff [*sic*!] ita privatus spiritu ut timoribus frigidus quasi actus et vita musculi in spiritu. *De motu locali* p. 94.

58. Palpitationes itaque fiunt a spiritu praeter naturum constituto .... *De motu locali* p. 96.

59. ... aliis animalibus spiritus tantum aliis fibra, caro, etc. *De motu locali* p. 96.

60. WH An sicut spiritus organum notus et sensus, ita caro cuius pars spiritus ist it tensus it motus. *De motu locali* pp. 86, 88.

61. Per arduam et altam scandentem timorem vi adhibet et defento spiritu minor aparet. *De motu locali* p. 56. Cf. ed cit. p. 78: Appetitus movet moto corporali videlicet spiritu....

62. ... quodcunque contrahit pro animae arbitrio habens in se spiritus motivus, cuius species musculus est. *De motu locali* p. 70.

63. WH An? nervus ut interventum judicis ... WH Nervi usus communicare sensibile cerebro ut fiat judicium. *De motu locali* p. 110. Cf. the long passage on p. 150 entitled Similitudo Analogica.

64. Cf. John Donley, Riolan and Harvey. *Annals of Medical History* 5(1923) 26-33.

65. Cf. Walter Pagel, *William Harvey's Biological Ideas.* New York, 1967 especially (Chapter VII) "Harvey's Vitalistic Criticism of Ancient Materialism" (pp. 251-282).

University of Kansas,
Lawrence, Kansas.

Can You Play a Game Without Knowing It?

by Bernard Suits

I

Let us suppose a person to be doing something which he suspects might be playing a game. (I am not begging the question at issue. I recognize that an advocate of the view that one cannot play a game without knowing he is playing it would refuse to acknowledge the sanity of such a suspicion. But it is not insane to suppose that someone —say an insane person— could have such a suspicion.) How would he set about trying to confirm or disprove his suspicion?

Consider the case of J.B. Lovegold. J.B. is a highly successful importer of gems. He is utterly without moral scruple, although he always conducts his business well within the letter of the law out of a keen sense of personal prudence. He is not certifiably insane nor does he suffer more than the usual amount of neurotic disturbance. He has a number of interests in life. These include the exercise of power, the satisfaction of his bodily appetites, the enjoyment of literature, music, and art, and the welfare and happiness of his family. He is not without hobbies, being a devoted philatelist. It is clear to him that he has an interest in increasing his monetary holdings, since money serves his other interests; and, since all of his appetites are enormous, and the exercise of power virtually limitless, he has an interest in increasing his monetary holdings without discernible limit. In short, subject to considerations of personal prudence and insofar as this does not interfere with his other interests, J.B. is out to make as much money as he can.

It is also the case that J.B. has for some time been losing some of his profits to another importer, one N. Croach. J.B. is aware of this fact. And it is the case that J.B. could force Croach out of the market (which would ruin Croach financially), since Croach's capital is very limited and J.B.'s is not. Furthermore, although this state of affairs has existed for more than a year, J.B. has not taken such obvious and safe steps to increase his profits. One day J.B. asks himself: "Why haven't I eliminated Croach? Is it because I have a personal regard for him? On the contrary, I have always found him detestable. I would be happy, as a matter of fact, to see him in the gutter. Am I concerned about his family, or his friends, who depend upon his power and good will? He is an orphan and a bachelor who heartily dislikes, and is disliked by, everyone with whom he comes into contact. There is absolutely no reason why I should not eliminate Croach and thus make more money. Furthermore, since I have known all these things for some time, why haven't I eliminated him? There must be *some* reason I persist in letting him remain in business. I seem to have set limits to the means I will employ for making money, yet there seems to be no reason why I have done this."

J.B. stops to mull this over, and arrives at the following solution: "I have made a small game of this business of making profits, or rather, I have willingly allowed a game to occur. I evidently prefer to have the fun of making as much money as I can with competition, even though this reduces my profits, rather than to make money exclusively as a means for serving my other interests. Is this irrational? Not at all. Just as I willingly make less money than I could because of possible dangers in adopting illegal means for making more, and I make less than I might because of the time I devote to my family and my stamp collection, so now I have hit upon a new interest. My work is now in part a game. But this game cannot continue if I seek to make more money by eliminating the other player. Well, well, to think that I have been playing a game for over a year without knowing it. Live and learn."

II

I submit the foregoing example as one instance of a person playing a game without knowing he is playing it. If this is accepted, then my task, or the major portion of it, would appear to be finished. But it may not be very readily accepted, and even if it is I propose to say somewhat more about unconscious game-playing. Prompted by some suggestive features of J.B.'s case, I shall attempt to develop the basis for a classification of ways in which one can play a game without knowing it. Following that I shall try to explain how some of these varieties of ignorance are possible.

But before I come to that perhaps I should deal in some way with the denial that what J.B. was doing was playing a game without knowing it. Such a denial might be made for one or both of two reasons. One might refuse to grant that what J.B. was doing was something of which he was not conscious, or one might refuse to grant that J.B. was playing a game, whether consciously or not. I shall not at this point consider the first of these objections; support for the contention that it is possible to perform intentional actions of which one is not conscious, and that that is what J.B. was doing, must await my later discussion of how it is possible to be ignorant of one's own game-playing. For the present I shall consider only the objection that J.B. was not playing a game *after* he reached the conclusion that that is what he was doing. And later on I shall set forth, and ask the reader to accept, a formal definition of game-playing. But I want to argue now that whether one agrees or disagrees that J.B. was playing a game (and, similarly, whether one accepts or does not accept the formal definition) is not essential to accepting the import of my argument. For if it is granted that one can unconsciously perform intentional actions, then those who maintain that one cannot play a game unconsciously must believe that there is something about games such that they, even though not all intentional actions, cannot be performed without being known to be performed. The burden of proof thus shifts to the objector. For if it is maintained that the case of J.B. is not a case of unconscious game-playing because J.B.'s actions, even when consciously performed, are not a game, then the objector must be able to show why their in fact being a game would preclude their being unconsciously performed. For the present, therefore, I shall continue to talk about the case of J.B. as though it were a case of unconscious game-playing.

III

Returning, then, to that example, let us ask just what it was that J.B. did not know when he did not know that he was playing a game. Evidently J.B.

knew what a game was (since otherwise he could not have come to identify what he was doing as a game), and he knew (short of knowing that it was a game, or course) what he was doing; among other things, he knew he was not eliminating Croach. He also knew that his continued permission of Croach's competition was not a means to some other end he wished to attain. Thus, although he knew what a game was, and he also knew what he was doing, he did not know that what he was doing was a game. How is such a thing possible? I suggest that it is possible in the same way in which it is possible to know the premisses of an argument without knowing the conclusion they entail. The premisses may not have been, so to speak, put together, but may exist separately in one's consciousness. There are two ways in which one can fail in such putting together of premisses. One can fail to be aware of the argument form whose premisses are instantiated by the unrelated pieces of knowledge, or, being aware of the argument form, one can fail to know that the unrelated pieces of knowledge instantiate its premisses.

Thus a person might fail to know that he is playing a game because he does not know what a game is, or, knowing what a game is, because he has failed to connect (in either of two ways) what he is doing with what a game is. He may, finally, not know what he is doing. In this case it is not a question of failing to know a definition, or of failing to make a logical connection, but of failing to be conscious of one's own intentions. Thus J.B., although he might know that he had intentionally ruled out eliminating Croach, might mistakenly believe his action to be for the sake of some end other than the end of playing a game. Or he might not even know he had made such a ruling. Or, knowing he had made such a ruling, he might not know he intended it. Finally, he might not know that his end is to make money.

Prompted by these consideration, I would like to make a formal presentation of a number of ways in which a person might fail to know he is playing a game he is in fact playing. In order to do this it will be useful to employ a definition of playing a game, and although I shall not insist that the definition is without loopholes, I shall hope that it is fairly convincing, and to that end I shall lead up to the definition by recounting the essential features of the J.B. episode.

J.B. knew that he was trying to increase his money. Then he discovered that although there were a number of ways in which he might do this, there was one way which he consistently rejected. He then realized that amking money was not the exclusive aim of his actions. He wished to increase his money, to be sure, but he wished to do this in ways which (apparently arbitrarily) excluded others. He thus came to see that his end was not simply to increase his money, but to do x: increase his money without using means m. And his reason for excluding m was solely so that he could do x. I submit that activities generally acknowledged to be games are subject to the same kind of analysis.

IV

I. *Definition.* To play a game is to engage in an activity directed toward bringing about a specific state of affairs, using only means permitted by specific rules, where the means permitted by the rules are more limited in scope than they would be in the absence of the rules, and where the sole reason for accepting such limitation of means is to make possible such activity.(1)

II. *Elucidation.* To the extent that the definition is adequate it states the necessary and sufficient conditions for anything's being a game.

III. *Stipulation.* The phrase "the sole reason for accepting such limitation of means is to make possible such activity" is (where X is a restriction of means to an end) equivalent in meaning to "x is intended not because it serves an external interest."

IV. *Elaboration.* If what agent A is doing conforms to the definition, then the following statements are true:

(1) A is seeking an end *e*.

(2) A has ruled out means *m* to end *e*.

(3) A intends (2).

(4) A intends (2) not because it serves an external interest.

(5) If (1) and (2) and (3) and (4), then A is playing a game.

(6) If A is playing a game, then (1) and (2) and (3) and (4).

V. *Assumptions.*

A1. If you know that you are playing a game G, then you know (i) that C are the necessary and sufficient conditions for anything to be a game, and you know (ii) that G fulfills C. (Thus, it is possible to fail to know you are playing a game either because you do not know what a game is, or because you do not know that what you are doing is playing a game.)

A2. If knowing "q" implies knowing "p implies q" and "p" then knowing "q" also implies knowing "If p implies q, and p, then q." (Thus, it is possible to know that a conditional is true, and to know that its antecedent is true, without knowing that its consequent is true, since you may not know that *modus ponens* is true.)

A3. If knowing "q" implies knowing "pr implies q" and "rp" and "If p implies q, and p, then q," then knowing "q" also implies knowing "rp is equivalent to pr." (Thus, it is possible to know that a conditional is true, and that a statement S is true, and that *modus ponens* is true, but not to know that the consequent of the conditional is true, since you may not know that S is the antecedent of the conditional. In the statement of A3, ignorance of commutation is intended to stand for the general case of ignorance that a known statement is the same as the antecedent of a known conditional.

VI. *Argument.*

If Agent A knows he is playing a game, the following statements are true.

(a) A knows (1) through (4). (A1,ii)

(b) A knows (5) and (6). (A1,i)

(c) A knows *modus ponens.* (A2)

(d) A knows that (1) through (4) affirms the antecedent of (5). (A3)

VII. *Conclusion.* On Assumptions A1, A2, and A3, if statements (1) through (6) are true, and if (a) or (b) or (c) or (d) is not, then A is playing a game and he does not know that he is playing a game.

V

Since (a) in the above list contains four items of information and (b) contains two, the list from (a) to (d) states eight necessary conditions for an agent's being conscious of the fact that he is playing a game. There are accordingly two hundred and fifty-five cases in which one or more of these conditions might not be met. However, some of these types of ignorance entail others; for example, ignorance of (a) entails ignorance of (3) and ignorance of (3) ignorance of (4). In terms of the proposed definition, therefore, a proper answer to the question whether there is any way unknowingly to play a game would appear to be that there are as many ways as there are non-redundant combinations of the eight kinds of ignorance listed. I propose now to illustrate four of these ways.

The cases I shall consider are all cases in which an agent is assumed to have knowledge of all of the items except one, since I am interested in illustrating one kind of ignorance at a time. Although there are eight such cases, I shall discuss only one each from among the four I have ordered with lower case letters, beginning with (d) and working back to (a).

Case (d). J.B.'s ignorance of the fact that he was playing a game evidently consisted in ignorance of (d). For he was not ignorant of what a game was, or of what he was doing, or of *modus ponens.* He was ignorant simply of the fact that what he was doing was playing a game. Put more generally, he failed to recognize that a thing was identical with itself in two separate occurrences. We may then ask how one can fail to recognize such an identity. It is not difficult to provide an illustration. I know Smith, but I did not know that the man who passed the door five minutes ago was Smith because the man who passed the door had a beard, and I did not know Smith had grown a beard. We might generalize and say that such ignorance is possible when the context of one occurrence of a thing differs from the context of another occurrence of the thing in such a way as to conceal the identity of the thing in its two occurrences. In the case of Smith, the relevant contexts are a hirsute and a non-hirsute context. In J.B.'s case he knew that such and such was what he was doing, and he knew that if one were doing such and such, then one was playing a game, but these two occurrences of his knowledge of such and such were not initially seen by him as occurrences of the same thing, since the different contexts of their occurrence masked their identity. The relevant contexts here might be the context of game-playing and the context of business. It is not at all unlikely that J.B. would fail to recognize that he was doing the same thing in his work that he was accustomed to doing only when he regarded himself as playing; e.g., in the context of a golf course or a poker table. On the other hand, it is not difficult to see how, upon reflection, he might come to make the identification.

Case (c). In entertaining the possibility that a person might be ignorant of *modus ponens* I do not mean that he has not had the opportunity to take a course in introductory logic, nor that he just happens never to have been conscious of using that argument form when he makes inferences of the kind logicians call *modus ponens.* I mean a person who is incapable of making such an inference. Imbeciles, or some kinds of imbeciles, might be examples of such persons. However, I do not wish to say that only imbeciles can be unconscious that they are playing games because of an unawareness of (c), since an imbecile who was unaware of (c) would probably also be unaware of (a), (b), and (d). Rather, I would like to call attention to what might be called an element in the psychopathology of everyday life. One need not be an imbecile to fail to make inferences of the form *modus ponens*; one might be suffering from what may be called logical aphasia, although this kind of aphasia is perhaps less generalized than the usual clinical types. Thus, I do not mean that a person so afflicted is never able to infer the consequent from a conditional together with its antecedent, but that this failure manifests itself on some occasions; specifically, on those occasions when the sufferer does not want to draw the conclusion entailed by premisses of which he is aware. Oedipus' protracted refusal to accept his true identity in spite of the evidence is a classical case in point. A game-player, too, might fear to know that he is playing a game. The case of an officer in combat playing with the lives of his men comes to mind.

Case (b). When someone is unaware of (b), he knows what he is doing, but he does not know that what he is doing is a game. This requires further explanation. If a person knows what he is doing, it might be objected, and what he is doing is playing a game, then surely he knows that he is playing a game. My answer is that by separating knowledge of (1) through (4) from knowledge of (5) I am able to say that he knows everything about what he is doing *except* that it is a game. It might still be objected that this exception to his knowledge is trivial, since it consists wholly in having failed to put a name to something: "...and, incidentally, people call such activity a game." This is simply the kind of knowledge, it might be argued, that a child acquires when he learns the names of things. But I am thinking of a person who is already aware of the word, and uses it correctly with respect to a variety of activities. Such a person calls x, y, and z games, but he does not know what there is about them that makes them games. Thus, when he comes to realize that what he is doing is playing a game, he has not merely applied a word to his actions, he has also learned that his actions are the same kind of thing he already calls game-playing. That this is new knowledge is evidenced by the fact that he is very likely to adopt a new attitude toward what he is doing, and that this new attitude is likely to issue in new behavior; e.g., ceasing to play what he now knows to be a game. This kind of ignorance, and its corrigibility, is also illustrated in other areas of our experience. The student did not think that an argument with contradictory premisses could be valid, but when he came to understand the meaning of validity he saw that it was. In game-playing, too, ignorance of the fact of such playing is possible where the player has reflected insufficiently on what it is to play a game.

Case (a). A person can be ignorant of (a) by being ignorant of (1), (2), (3), or (4), or of any combination of these. I shall consider only the case in which he is ignorant of (4). In this case the person knows he is seeking a certain end, and he knows that he has intentionally ruled out certain means to this end, but he does not know that he has done the latter solely for the sake of the activity in question. Perhaps the most plausible case is the case in which his unawareness of the real reason he has ruled out some means is due to his mistaken belief that he has ruled out those means for a different reason.

Thus, although J.B. has ruled out the elimination of Croach, he may mistakenly believe his reason for doing this was because there is some ambiguity in the laws governing the conduct of his business such that the elimination of Croach might result in J.B.'s becoming involved in a costly lawsuit. The issue is not whether there is or is not such an ambiguity in the law, nor simply whether J.B. believes there to be, but whether he believes himself to be acting on the basis of such a belief. How is this possible? The most plausible case is the case of self-deception. A J.B. less honest and open with himself than the J.B. of the original example might not want to know that he is playing a game with something he regards with as much gravity as he regards his life's work. He might therefore keep this knowledge from himself by inventing a reason for permitting Croach to remain in business. He might mask his own intentions from himself by means of the fiction that it would not be safe to ruin Croach.

VI

Assuming that the possibility of unconscious game-playing has been made to some degree convincing, one might question whether this fact is more than a mildly interesting curiosity. If playing a game without knowing it were merely a logical possibility, then the bare knowledge that such a possibility existed would perhaps not be very important. But if it is at all likely that very many of the things we do are games in disguise, then I should think it would be very important indeed to suspect that this might be the case. If we were to discover certain of the things we do to be game-playing, we would be very likely to alter our conduct in one of two directions. For the discovery that one is playing a game would be likely to result in either chagrin or delight, depending on the player's attitude toward games and toward the proper conduct of his life. The first response would probably be to stop playing the second to play better.

A somewhat more grand —perhaps grandiose— defense of the importance of the thesis at issue may be made. If one is interested in inquiring into "the meaning of life" the possibility of unconscious game-playing, it seems to me, removes the chief obstacle to taking seriously the possibility that life is a game in which all of us are players.(2)

NOTES

1. The author has argued for the adequacy of this definition in "What Is A Game? " *Philosophy of Science*, Vol. 34, No. 2, June, 1967

2. See the author's "Is Life A Game We Are Playing? " *Ethics*, Vol. 77, No. 3, April, 1967

University of Waterloo,
Waterloo, Ontario.

A Third Century Taoist Treatise on the Nourishment of Life: Hsi K'ang and his *Yang-sheng lun.*

by Jerry Swanson

After the death of the great Chinese general Ts'ao Ts'ao, (A.D. 155-220) in A.D. 220, the Wei dynasty which he founded was slowly undermined by the power of the rising Ssu-Ma clan. Ts'ao Ts'ao's successor was his son Ts'ao P'ei (A.D. 188-227), but since Ts'ao P'ei had no particular military talents, real power soon fell into the hands of the prime minister Ssu-Ma I (A.D. 178-251). Upon his death this position fell to Ssu-Ma Shih. The Ts'ao clan tried several times to reassert their power, but were defeated by Ssu-Ma I in A.D. 251 and Ssu-Ma Shih in 255. When Ssu-Ma Shih died in 255 he was succeeded by his brother Ssu-Ma Chao (A.D. 211-265). In A.D. 260 Ssu-Ma Chih put down an intrigue by the Emperor Ts'ao Mao by ruthlessly killing him and placing a puppet on the throne. Finally in A.D. 265 another Ssu-Ma,Ssu-Ma Yen (A.D. 236-290) threw out the Ts'ao clan altogether and proclaimed himself the first emperor of the Chin dynasty.

Despite the deadly behind-the-scene struggles between the Ts'ao and Ssu-Ma clans, the Chinese capital of Loyang was the scene of great literary productivity. The Buddhists were represented by Dharmakala(1) (fl. A.D. 250) and Sanghavarman(2) (fl. A.D. 250) from India as well as the Chinese priest Po Yen (fl. A.D. 250). The Ts'ao clan and Ssu-Ma clans themselves produced a considerable amount of literature which has survived to the present day. Ts'ao Chih, (A.D. 192-232), Ts'ao P'ei's younger brother became known in later generations as one of China's greatest literary figures. Ssu-Ma I and Ssu-Ma Chao also enjoyed something of a reputation as writers during their time. There were many other literati living at Loyang at this time such as the poets Sun Kai (d. A.D. 261) and Ying Chu (A.D. 190-252), the scholars Huan Fan (d. A.D. 249) and P'ei Hsiu (A.D. 224-270) —and the Seven Sages of the Bamboo Grove.

As a consequence of the struggles between the Ts'ao and Ssu-Ma clans many literati were drawn into the battle simply because they were friends or favorites of the contestants. When a man fell into disgrace during these times he was often sentenced to death along with his entire family down to the third generation and most of his known friends. For this reason it became very expedient to remain silent and have as few political associations as possible. However, there is a danger in generalizing that this political situation reflected itself in the literature and philosophy of the times.(3) To be sure there was a certain retreat into the nihilism of Taoism by the literati, but there were also many staunch Confucians. Quite often the inability to express oneself within the political sphere simply produced a retirement from the world by the refusal to accept office.

In such an uncertain atmosphere, then, lived Hsi K'ang and the Seven Sages of the Bamboo Grove. Historical details about Hsi K'ang the man are difficult to piece together. He seems to have been born in the *Kung-yüan* era,

A.D. 223 and died in A.D. 262. His family was originally from a town called Shang-yü in present day Chekiang Province. Fleeing enemies, his family moved to a small village called Chih in modern Anhui Province, where they changed their name from the original Chi, to Hsi, the name of a nearby mountain. Despite these humble circumstances Hsi K'ang was able to marry into the Imperial Ts'ao clan and accept a titulary post at Shan-yang,(4) in the present Ch'in-yang hsien slightly north of the capital of Loyang. It was during these years that he started meetings of a small band of Taoist worthies who met from time to time in a bamboo grove on his property.(5) These men became known later as the "Seven Sages of the Bamboo Grove." The other members of this "magnificent seven" were Juan Chi (A.D. 210-263),(6) Juan Hsien (fl. A.D. 260),(7) Wang Jung (A.D. 234-305),(8) Liu Ling (A.D. 221-300),(9) Shan T'ao (A.D. 205-283),(10) and Hsiang Hsiu (A.D. 221-300).(11) Although now a member of the Imperial clan Hsi K'ang would have nothing to do with palace politics. Shan-yang was something of a center of Taoism of the time and Hsi K'ang devoted all of his time to its study. In a famous letter to his friend Shan T'ao refusing Shan's recommendation of him to office, Hsi said:

> Besides, I have studied in the esoteric lore of the Taoist masters, where a man's life can be indefinitely prolonged through eating herbs, and I firmly believe this to be so. To wander among the hills and streams, observing fish and birds, is something I would have to give up forthwith. Why should I relinquish what gives me pleasure for something that fills me with dread? (12)

And again:

> Of late I have been studying the techniques of prolonging one's life, casting out all ideas of fame and glory, eliminating tastes, and letting my mind wander in stillness: what is most worthwhile to me is Inaction.(13)

Hsi K'ang deliberately slighted a favorite of Ssu-Ma Chao, a certain Chung Hui (A.D. 225-264), when Chung came to visit him seeking advice on a piece of amateur philosophy he was writing.(14) For this he was later implicated in a flase plot to overthrow the Ssu-Ma clan and was sentenced to death.(15) Hsi K'ang certainly understood the eventual consequences of his insult to Chung Hui, but here again his letter to his old friend Shan T'ao gives us some insight into his personality:

> One can be so constituted that there are things one cannot endure; honest endorsement cannot be forced. So it is perhaps idle to talk about the familiar "man of understanding" who can put up with anything, who takes no exception to vulgarity about him, but who still preserves his integrity within; who goes along with the vacillations of the times without ever feeling a twinge of regret. Lao Tzu and Chuang Chou are my masters: They held mean positions.(16)

Before his execution he wrote some verses called "Black Exasperation", *Yu-fen*, (17) expressing his inner feelings, but outwardly he remained calm; when his family came to the prison in his last hour Hsi K'ang asked for his lute and played the melody *T'ai p'ing* "Great Peace." When he had finished it he sighed and said: "Now the melody T'ai p'ing dies with me."(18)

Many years later Hsiang Hsiu, another of the Seven Sages, wrote his *Ssu chiu-fu* "Poetical Essay on an Old Friendship" in which he said:

> Hsi K'ang excelled in all the arts, and especially was a wonderful musician. When his last hour drew near, looking at the setting sun he drew his lute unto him and pulled the strings. At that time I travelled westwards, but now returning I again pass where he used to live. Now night is falling, the darkness brings a piercing cold. Nearby a man plays that flute, its tones are sad and forlorn. They make me think of our friendship in the old days when we passed the time together...(19)

The *Yang-sheng lun* belongs to the literary genre called *lun*, translated indiscriminately as discourse, disquisition, essay or treatise. Its origin is probably from the *Book of Changes.*(20) We know that *lun* as a genre was well established by Han times, but the first extant piece of literary criticism on the *lun* is by Emperor Wen (Ts'ao P'ei) of the Wei dynasty:

> Literature as a whole comes from the same root, but in its lesser details it displays great difference. Memorials and imperial decisions should be elegant, letters and discussions *(lun)* should be well reasoned:
>
> Memorial tablets and eulogies of the dead should stick to facts, lyrical poetry and *fu* require to have symmetry. These four classes being different, the result is that those who have ability with any one of them are lopsided, and only those with comprehensive talents can perfect their style.(21)

Liu Hsieh (A.D. 465-522) in his *Wen-hsin tiao-lung*, or *The Literary Mind and the Carving of Dragons*, a later more extensive treatise of Chinese literary criticism discusses the genre as follows:

> The principles propounded by the Sage are known as *Ching*, or Classics, and the works which explain the Classics and set forth their underlying ideas are known as *lun*, or treatise or discourse. *Lun*, or to discourse, literally means *lun*, or to set in order.
>
> ***Lun*** means to take into consideration a variety of statements for the purpose of examining minutely a specific idea.(22)

Speaking of the genre during the time of Hsi-K'ang, Liu says:

> During the period when the Wei first came into power, logic and law were the studies of the day. Fu Ku (209-255) and Wang Ts'an were experts in the logical analysis of name and reason. By the time of Cheng-shih (240-248), literature became the first concern. It was Ho Yen (? —249) and his group, however who were responsible for the beginning the writing of the metaphysical treatises. At this time, Tan (or Laotzu) and Chou (or Chuangtzu) came into focus and competed with Ni-fu (or Confucius) for the highway. The "Ts'ai-hsing" of Lan-shih (or Fu Ku), the "Ch'ü-fa" of Chung-hsüan (or Wang Ts'an), the analysis of sound by Shu-yeh (or Hsi K'ang), [a treatise on the theme that the human voice has no intrinsic quality expressing sadness or happiness] .... are all the results of profound and unique insights, original with each author, and all are characterized by penetrating and sparkling arguments creating logic that is close-knit and flawless. These authors were indeed men of outstanding talents.(23)

It is against this background that we must approach Hsi K'ang's *Yang-sheng lun.* We must account for, as nearly as possible, the classics or classic which Hsi K'ang's *Yang-sheng lun* attempts to explain, and understand, as well as possible, the underlying ideas he considered these classics to have.

The term "*Yang-sheng*" in the title of Hsi K'ang's *lun* is most certainly a technical term of Taoist philosophy. In ordinary usage it meant, and still means, "to nourish the living" or "to nourish life" depending upon the context. It is used in this sense in an early passage in Mencius: "... to nourish the living and bury the dead without remorse is the beginning of the Way of Kings."(24)

In a Taoist context the first occurrence of *Yang-sheng* is in the *Chuang tzu* where it appears as the title of one of the books, the *Yang-sheng chu.* (25) The book is divided into four parts; the first part sets out the general principle:

> There is a limit to our life, but to knowledge there is no limit. With what is limited to pursue after what is unlimited is a perilous thing; and when, knowing this, we still seek the increase of our knowledge, the peril cannot be averted. There should not be the practice of what is good with any thought of the fame (which it will bring), nor of what is evil with any approximation to the punishment (which it will incur): —an accordance with the Central Element (of our nature) is the regular way to preserve the body, to maintain the life, to nourish our parents, and to complete our term of years.(26)

The second part is the famous parable of the Master Butcher. Upon hearing the butcher describe his art the king said: "excellent! I have heard the words of my cook, and learned from them the nourishment of (our) life."(27)

The concept of *Yang-sheng* in the *Huai nan-tzu*, a compendium of natural philosophy compiled by a group of Taoist worthies supported by Liu An, the prince of Huai nan, is also a technical term. *Yang-sheng* is mentioned in three places; the most important instance may be translated:

> Therefore if one knows the greatness of the universe, then he cannot be harassed by life and death. If one knows the harmony of *Yang-sheng*, then he cannot be in dire straits in the world.(28)

The commentator, Hsü shen, explains the harmony of *Yang-sheng* as cheng tao, or the "proper Tao." And that if one cultivates the "proper Tao," one cannot be shown to be powerless in the empire. This adds a slight Legalistic note to the text.

During the *Ch'un-ch'iu* period (722-481 B.C.) the notion of immortality began to appear in Chinese literature. Before that time the ideal of the people was to live on the earth to an advanced age and die a natural death. Those who held this new conception of immortality desired to escape the world altogether and live in the high mountains as a *hsien*, or immortal.(29) The earliest description of a *hsien* appears in the *Chuang tzu:*

> ... Far away on the hill of Ku-shih there dwelt a Spirit-like man whose flesh and skin were (smooth) as ice and (white) as snow; that his manner was elegant and delicate as that of a virgin; that he did not eat any of the five grains, but inhaled the wind and drank the dew; that he mounted on the clouds, drove along the flying dragons, rambling and enjoying himself beyond the four seas; that by the concentration of his

spirit-like powers he could save men from disease and pestilence, and secure every year a plentiful harvest.(30)

During Ch'in and Han times the concept of a *hsien* became popularized and many people devoted their energies to developing drugs and potions which would enable them to become *hsien*. An important passage in the *Han Shu*, the history of the Han dynasty, records the situation:

> When Ch'in Shih Huang first unified the empire, he indulged in the cult of *hsien* immortality. Thereupon, he sent people like Hsü Fu and Han Chung to sea, with unmarried boys and girls, in search of *hsien* as well as drugs. But (these people) took the opportunity to run away and never came back. Such efforts aroused the resentment and hatred of all under heaven. With the rise of Han, Hsin-yüan p'ing of Chao (in the reign of Wen-ti, 179-157 B.C.), Shao-weng, Kung-sun Ch'ing, Luan Ta, and others of Ch'i, all received honors and favors from Emperor Wu-ti on account of their (claimed magical powers, such as) acquaintanceship with *hsien*, alchemy, sacrificial offerings, serving and controlling spirits, and going to sea to search for *hsien*, and drugs. Gifts bestowed on them amounted to thousands of measures of gold. (Luan) Ta was particularly honored and even married a princess. Titles and positions were heaped on him to such an extent that all within the four seas were shocked. Therefore, during the Yüan-ting and Yüan-feng periods (116-103 B.C.), there were thousands of *fang-shih* in the areas of Yen and Ch'i who protruded their eyes and clasped their hands, saying that they knew such arts as those of achieving *hsien* immortality, of sacrificial offerings, and of obtaining blessings.(31)

Yü Ying-shih argues quite convincingly that Huang Ti, or the Yellow Emperor was attached to the sage Lao-tzu to form a new school of Huang-Lao Taoism.(32) The Bibliographical treatise of the *Han Shu*, or dynastic history of the former Han, lists ten works extant in Han times under the school of *hsien* Immortalists *(shen hsien chia)* and four of these are attributed to Huang Ti.(33) Huang-Lao Taoism evidently taught that man could, with the proper instruction of *fang shih*, or Taoist magicians, achieve physical immortality.

Thus we can see the traditions with which Hsi K'ang were working were not at all new. Also during Later Han times there arose a pharmaceutical-botanical literature associated with the names of Huang Ti and Shen Nung. Of the 260 odd works listed in the *I-fang*, or pharmaceutical subsection of the Bibliographical Treatise of the Sui dynastic history (compiled in A.D. 641, but probably based on earlier catalogues(34)) ten were either attributed to Huang Ti or Shen Nung or had their names in the titles.(35) Eleven titles have the term immortal or divine immortal in them or are obviously alchemistic in nature; an example: *The Classic of Miscellaneous Prescriptions for Immortality*. There was also a *Yang-sheng* literature: *Important Techniques in the Nourishment of Life; The Classic on the Nourishment of the* (physical) *Body; Dietary Prohibitions in the Nourishment of Life; The Chronicle (chüan) of the Nourishment of Life; Important Prescriptions for the Nourishment of Life for the Emperor and King.*

We might well ask what relevance a mere book title which appears in the Imperial Collection in the middle of the seventh century has for third century

Taoism? Hsi K'ang's *Yang-sheng lun* has been preserved in various editions, the best known of which and the one used for this paper is the Wen hsüan (hereafter WH) a famous anthology of Chinese literature compiled by prince Hsiao T'ung (A.D. 501-531) of the Liang dynasty. The WH enjoyed a great popularity and in A.D. 660 a commentary was written on it by Li Shan (d. A.D. 689). Li Shan spotted a number of passages in Hsi K'ang's *lun* which came directly from the *Chuang tzu, Huai nan tzu, and the Nourishment of Life Classic* and *The Essentials of the Nourishment of Life* of the *Yang-sheng* literature.(36) These quotations form a two way street. From them we can get some idea of the content of the *Yang-sheng* literature prior to Hsi K'ang as well as understand in what ways they influenced his own thought.

Prior to its use by Hsi K'ang, the technical term, *Yang-sheng*, changed in meaning. In the *Chuang tzu* it served as an admonition to act within natural limits. In the *Huai nan tzu* it was closely related with the concept of harmony. And in the hands of the Huang-Lao school it was bound up with the notion of immortality. All of these meanings find a place in Hsi K'ang. While the *Yang-sheng lun* is certainly an important essay in the history of Taoism, it will probably strike the western reader as an essay having little to do with philosophy. One reason for this impression is that in their philosophies the Taoists are accustomed to acting in accord with all of nature rather than just human nature. They try to unite things we never think of trying to unite; they sense that in certain subtle ways right action must relate to things like the silence of the stars and the cold wind of a mountain summit. If a man is disposed to act in harmony with all of nature, if he is disposed to choose the mean writ large rather than man's mean, then he deserves the name *Hsien.*

On the Nourishment of Life(37)

For generations there have been some who have said that immortality(38) could be obtained by study and that "no death"(39) could be attained by force. Others have said that the highest longevity was one hundred and twenty years(40) and this was the same both in the past and in the present. Cases of living beyond this point are either fabulous or false. But these two opinions miss the point. Allow me to try to roughly discuss it.

Now the immortals, although not seen by the eyes, are however recorded in the annals and are transmitted by the former histories.(41) Comparing these sources and discussing them, their existence should be certain. It appears that the special endowment of this extraordinary breath(42) is received from nature and not reached by accumulated learning. As to the nourishment of life, if you acquire the principle of it in order to exhaust the potential of your lifespan, then the oldest can attain to one thousand years while the youngest can count a hundred years;(44) that might be possible, but still in the world there are none who have refined it and therefore there are none who can achieve it.

Why do I say this? When taking drugs for sweating, sometimes it doesn't work, but when the emotion of shame is concentrated, sweat freely pours forth.(45) If one hasn't eaten for a day, then he longingly thinks of food. However, Tseng tzu(46) in dwelling in mourning for his mother did not eat for seven days. If at midnight one sits in meditation, then he becomes drowsy and thinks of sleep. while if he inwardly harbours great anxieties then, at the arrival of dawn, he has not thought of sleep. Only a stiff brush can dress the hair, only strong wine brings forth color; with these means alone can such results be acquired. Only the anger of a brave man produces in him a strange looking flush, and makes his hair stand on end, pushing up his cap.(47) From this we can say the relationship of the spirit to the body is like the state to a ruler; when the spirit is irritated on the inside and the body withers on the outside, it is like when the sovereign is confused above and the state is in disorder below.

Now when they sowed grain in the time of T'ang,(48) it so happened that the grain received the benefit of a single watering and, although it became hot and dry again, the grain that had the water died last. Thus the benefit of one watering certainly cannot be denied. And still the people of the world commonly say one fit of anger is not sufficient to undermine one's nature, one instance of grief is not sufficient to injure one's body. This is taking things too lightly and dismissing them; this is like not recognizing the benefits of one watering while hoping for good grain from parched seedlings. Thus the superior man knows that the body relies upon the spirit to stand; the spirit relies upon the body in order to exist. Apprehending that the principle of life is easy to lose, the gentleman knows the harmfulness to life of a single mistake. Therefore he should cultivate his nature in order to protect the spirit, and pacify the inner heart in order to keep the body intact. Love and hate should not nest in the emotions; anxiety and joy should not remain in the mind. Thus tranquil and desireless the body and breath are concordant and peaceful; moreover proper breathing regulates the intake and outgo;(49) proper eating nourishes the body(50) causing it and the spirit to be close, and the inside and outside to be in accord.

Now, as for those who cultivate the fields, ten *hu*(51) from one acre is called an excellent field; this is a general estimate in all the world. What is not known is that with a *ch'ü* field one can get over one hundred *hu*.(52) Fields and planting are the same, but so far as planting and nurturing methods are different, the results will differ greatly from one another. To say a merchant may have no tenfold profit and the farmer has no hope for a hundred *hu* harvest —this is a viewpoint of inflexibility. That beans cause one to gain weight, the elm tonic(53) causes one to fall asleep, the *ho huan*(54) dispels anger, and the *hsuan* herb(55) causes one to forget sorrow — the stupid and the wise both know this. Strong leeks hurt the eyes, globe fish(56) do not nourish —everyone knows this. Lice, while living on the head, turn black; the musk deer because of chewing ginko becomes fragrant. The necks of people who reside in difficult environments develop boils.(57) The teeth of people who live in Chin become black.(58) Inferring from this, the essence of whatever we eat stains our basic character and dyes our body. Everything has a mutual interaction. Why should it only be that steaming causes heaviness and doesn't cause lightness; why should damage cause only blindness and not clear vision; fragrant plants cause yellowing and not durability; perfume cause fragrance and not prolong life? Thus it was that Shen Nung(59) said: "The higher class of drugs nourish fate, the middle drugs nourish one's nature."(60)

He well understood the principle of fate and nature because of his being conversant with (the principle of) nourishment.(61)

However, the people of the world do not observe this; only the five grains(62) they watch, only to sounds and colors are they addicted. Their ears indulge only in lewd music; the various flavors of things burn their innards — sweet wine boils their stomachs. Fragrance rots their bones and marrow; joy and anger confuse their regular breathing; thoughts and plans waste their spirit; grief and happiness injure their balance and purity. Such a frail body —there is more than one way of causing harm to it. With a body so easily exhausted it encounters enemies both internally and externally; the body is not made of wood or stone, how can it last for long? Those who use themselves in an extreme manner drink and eat without restraint: their body has one hundred diseases; they indulge in sex without tiring to the point that their life is cut short; through the misery of wind and cold, and through the injury of the hundred poisons, in the course of things, they perish young because of these myriad difficulties. The people of the world all know and ridicule and pity them, saying they are not good at maintaining life. As for controlling the body, if one neglects the principle, he loses by trifles. Accumulated trifles become injury, accumulated injury brings decline. Following the deterioration of health, one becomes white haired, following white hair, one gets old; following old age, one dies. Stupidly they do not know the reason; the people below middle intelligence call it nature.(63)

Suppose there are some who are a little enlightened; lamenting and hating at the beginning of difficulties, they do not know to take precaution against the numberless dangers before their omens have appeared. Therefore in the time of Duke Huan(64) when he contracted a mortal disease, he was angry because Pien Ch'üeh(65) saw it early. This is because the day he felt pain he regarded as the start of the sickness. Harm comes from a mere trifle; the attempts to cure it come only when it is manifest. That is why there are unsuccessful cures. Because one has galloped through the realm of ordinary men, one has the concept of an ordinary lifespan; looking up and looking down all men are like that. By means of the many, he proves it to himself and by means of similarity he consoles himself thereby saying: "Such is a principle of heaven and earth —simply this and nothing more! "

Suppose he hears of the nourishment of life,(66) then he judges it with his own opinions and calls it untrue. Another, although perhaps less doubtful, has some hope for it, but he doesn't know what it is all about. Another may devote himself to taking drugs —if after six months a year there is no result, his will declines and in mid-course he gives it up. Others try to increase their vitality gradually as if they were using an irrigation ditch to drain the sea at Wei-lü.(67) Still another controls his emotions and desires, cutting away glorious aspirations, but still lusts after that which is constantly before his eyes. What they hope for is found only in the distant future. Moreover fearing in his bosom the loss of both, he is timid; the heart fights from within, things entice from without; in their exchange they bring each other to fall —in this way it will come to nothing.

Now the subtlety of the highest things can be known by reason, but it is difficult to understand with the eyes alone; it is like the *Yu-chang* (68) tree, only after seven years of growth can it be distinguished from another similar tree. To traverse the road of non-attachment and quietude by means of a quarrelsome heart, the thought of this comes quickly, but the reality is late. He hopes for it to be near, but the result is far away. Therefore one cannot get to the desired end. As to those with fixed aims,(69) because they have no results, they pursue it no further while those who do pursue it, lose out because they do not pursue it with a single heart. Those who are inclined toward only one method, because of the fact that they are not comprehensive, achieve nothing. Those who pursue it as technique drown themselves in petty methods. They are all like this; therefore, of those who desire it, not one in ten thousand can accomplish it.

If one is good at nourishing life, then he is not thus! He is pure and void, tranquil and expansive;(70) his ego is small and his desires few. Knowing that a reputation and rank injures power, he therefore takes them lightly and makes no plans for them. This is not a case of first desiring something and then later suppressing it. Knowing that good eating harms the basic nature, he therefore casts it out and does not care for it. This is not being greedy while later suppressing it. Because external affairs bind the mind, he doesn't dwell on them. Spiritual power, because it is pure and pristine, is alone manifest in him. He is expansive and has no anxieties nor crises; he is still; he has no thoughts nor worries. Moreover he protects himself with the Unity; he nourishes himself with harmony.(71) The principle of this harmony daily becomes more apparent and he is in tune with the Great Concord.(72) Only then can he steam himself with the ling-chih plant(73) and drink the unfermented wine,(74) only then can he pacify himself with the five musical strings of the zither.(75) He does nothing, yet is self content.(76) His body is wonderful; his mind mysterious.(77) He loses himself in joy, yet later his happiness is sufficient. He abandons life, while in the end the body is preserved. Proceeding like this he can easily compare in age to Hsien men(78) and quarrel about years with Wang Tzu-chiao.(79) How can it be that he does not exist?

Those Warring State philosophers who have been called "Taoist" since Han times were in part concerned with expanding the range of meaning of certain Confucian ideas which they inherited and to which they stood in reaction. While the Confucians were concerned with human knowledge, the Taoists had an eye to what was beyond man. While the Confucians were concerned with ritual, ranks, and the maintenance of the proper time of the rhythms of society, the Taoists had an ear to the sounds of Nature. What was noble and ignoble in the human sphere had no significance in the K'un Lun mountains of the Taoist ideal. Although there was probably much more agreement between the two types of philosophies, the Taoist texts seem to offer a complete philosophical package which stands in reaction to the Confucian at almost every point.

Another aspect of Taoism, the shaman and alchemical traditions were also important, and came into its own with Ch'in Shih Huang Ti, China's first emperor. The growing interest in "pills of immortality" and drugs in general informs later Taoism with what was regarded as a *potential* method for achieving the ideals mentioned in the earlier texts. Thus Hsi K'ang lived at a time when a general synthesis could be made between these two aspects of Taoism. Whether he thought of the *Yang-sheng lun* as attempting that synthesis or whether he thought it already accomplished, it is clear that he stands at a time in the history of Taoism when fang-shih alchemical traditions became incorporated into the strictly philosophical literature. It is this synthesis with the alchemical *fang-shih* that prevents the *Yang-sheng lun* from being a mere rhapsody on a theme of Chuang Chou. Many of the rhetorical devices and methods of argumentation are taken from Chuang Tzu, but the base is not for the older ideals themselves, but for the older ideals achieved in a new way.

Hsi K'ang was a man who lived the philosophy he believed and like philosophers everywhere, tried to justify his ideals in his philosophical prose. It was through his philosophical prose and not his experiments with drugs that he was "pure and void, tranquil and expansive;..."

NOTES

1. His Chinese name was T'an-ko-chia-lo; see E. Zürcher, *Buddhist Conquest of China* (Leiden, 1959), p. 55. I wish to thank Professors Hellmut Wilhelm and Y.S. Lao, David Knechtges, Jack Dull and Mr. John Hughes for assistance on various parts of this paper.

2. There is some question whether this was another name for the Sogdian K'ang Seng'k'ai. He and Dharmakala as well as the two Parthians T'an (wu) -ti and An Fa-hsien arrived at Loyang shortly after A.D. 250. See Zürcher, *Conquest*, p. 55.

3. See E. Balazs, "Nihilistic Revolt or Mystical Escapism," *Chinese Civilization and Bureaucracy; Variations On a Theme*, H. M. Wright trans. (New Haven, 1964), p. 226-254. Balazs seems to carry this thesis too far. P'ei Wei (A.D. 267-300), although he lived a generation later than the Seven Sages of the Bamboo Grove, was a staunch Confucian and reacted to the chaos of his times by dogmatically reasserting Confucian values.

4. His rank was *Chung-san-ta-fu*, a sinecure position which allowed Hsi K'ang time to study and write. In his letter to Shan T'ao he refused an office of higher stature which would have meant actual administrative control. See note 12.

5. Professor John C. Ferguson has pointed out that *Chu-lin* in the phrase *chu-lin ch'i-hsien*, or literally "bamboo grove —seven sages (worthies)" is probably not "bamboo grove" but an actual geographical location. See R. H. Van Gulik, *Hsi K'ang and His Poetical Essay On the Lute*, Monumenta Nipponica (Tokyo, 1941), p. 1, n. 1.

6. Of the Seven Sages probably the most famous were Hsi K'ang, Hsiang Hsiu, and Juan Chi (sometimes spelled Yüan Chi). Juan Chi's biography in the Chin dynastic history describes him as:

"Handsome and imposing in appearance, and of open and far ranging mind. Juan Chi was fiercely independent and did exactly as he pleased. Neither joy nor anger was ever expressed on his face. Sometimes he stayed shut up in his room studying books for several months on end, without ever going out; sometimes he went up into the mountains or to the waterside, forgetting for several days to return home. He read an enormous amount and had a special love for Chuang-tzu and Lao-tzu. He was a great drinker, an accomplished performer on the lute, and skilled in whistling. While he was following out a train of thought, he would quite forget about the outside world. His contemporaries often called him a madman."

(Balazs, p. 236-37.) Juan Chi was a great drinker to be sure and one tradition has it that he accepted a certain administrative post because the office had a good wine cellar. To escape a forced marriage to one of the Ssu-Ma's daughters, he remained dead drunk for sixty days. He was also a great poet and prose writer; in one of his more serious moments he satirized the Confucians, comparing them with the lice that inhabit the trousers:

"Have you ever seen the lice that inhabit a pair of trousers? They jump into the depths of the seams, hiding themselves in the cotton wadding, and believe they have a pleasant place to live. Walking, they do not risk going beyond the edge of the seam; moving, they are careful not to emerge from the trouser leg; and they think they have kept to the rules of etiquette. But when the trousers are ironed, the flames invade the hills, the fire spreads, the villages are set on fire and the towns burned down; then the lice that inhabit the trousers cannot escape.

"What difference is there between the gentleman who lives within a narrow world and the lice that inhabit trouser legs? " (Balasz, 238.)

7. Juan Hsien was a nephew of Juan Chi and supported his uncle for a time. Hsien was also very fond of drinking and carousing. He was reported to have had an affair with one of his aunt's maidservants and had an illegitimate son by her. Like Juan Chi and Hsi K'ang, he was an excellent poet and musician.

8. Wang Jung was always late to the meetings of the Seven Sages which made Juan Chi quite angry. On one occasion Juan Chi said "Is this boor once again going to wreck our train of thought? " Wang overhearing replied, "Is it all that easy to wreck your train of thought, Sir? " (Balazs, 248n.) Wang Jung was a very wealthy man who was at the same time very miserly. He went so far as to remove the pits from his plums so someone could not sell the same variety. It is interesting to observe that someone with Wang's money could still have been a Taoist and respected by his friends enough to be one of this famous group.

9. Although all of the Seven Sages were known for their dinking ability the one traditionally held to be champion of them all was Liu Ling. He was supposedly followed by his servant with a shovel with orders to bury him where he fell if he died; another servant carried an extra jug of wine lest he should run out. Even today alcoholics in China are referred to as spiritual descendents of Lui Ling. See T. K. Chuan, "Yüan Chi and His Circle," *T'ien Hsia Monthly*, IX (1939), 472-477. Although Liu was well known as a poet the only surviving piece is a poem eulogizing wine, the *Chiu-te-sung* which has been translated by G. Margoulies in *Le Kou-wen Chinois* (Paris, 1926), p. 124. According to one anecdote Liu Ling was totally naked when suddenly visited by some friends. They were startled but he calmly replied: "Heaven and earth are my dwelling and this room is my breeches. Who asked you, gentlemen, to come inside my breeches? " (Balazs, 239.)

10. Shan T'ao had great respect for Hsi K'ang and after the latter's death provided an education for Hsi K'ang's two children. Shan sided with the Ssu-Ma clan in their struggle with the Ts'ao clan, but also tried to keep Hsi K'ang's friendship. Hsi K'ang broke off relations with Shan in a famous letter. Shan and Wang Jung who sided with the Ssu-Ma rose to be high officials in the Chin dynasty.

11. Hsiang Hsiu was a great student of Taoism and wrote a special study of the work called the *Chuang tzu.* According to one tradition his commentary was stolen after his death and published as his own by a certain Kuo Hsiang. The work survives and today is known as the *Hsiang Hsiu-Kuo Hsiang Commentary.* Hsiang wrote a criticism to Hsi K'ang's *Yang-sheng lun* called *Nan yang-sheng lun,* or "Difficulties with the *Yang-sheng lun.*" Hsi K'ang replied with an answer, *Ta nan yang-sheng lun,* or "Replies to Difficulties with the *Yang-sheng lun.*" Hsiang Hsiu appears to give a hedonistic interpretation of *Yang-sheng* in his essay "Difficulties"; this seems to be in accord with a rival school of Taoism stimulated by the Yang Chu chapter of the *Lieh Tzu.* A.C. Graham, *The Book of Lieh Tzu* (London, 1960), 143 note: "The phrase yang-sheng ('tending life,' 'tending the living') had different meanings for different schools. For individualists of the 4th century B.C. (deriving from the historical Yang Chu himself) it meant the satisfaction of personal needs without injuring health and life. For Confu-

cians, 'tending the living' and 'taking leave of the dead' were the filial duties of supporting and decently burying one's parents. The present passage gives the Confucian terms a hedonist reinterpretation."

12. J.R. Hightower trans., "Hsi K'ang: Letter to Shan T'ao," *Anthology of Chinese Literature; From Early Times to the Fourteenth Century*, Cyril Birch ed. (New York, 1965), 165.

13. Hightower, 165.

14. This incident is recorded in the *Shih-Shuo Hsin Yü*, or *New Account of Tales of the World:* "When Chung Hui had barely finished editing the 'Treatise on the Four Basic Relations between Natural Ability and Human Nature' *(Ssu-pen lun)*, he wanted very much to have Hsi K'ang look it over. Putting the manuscript in his bosom, he went to the latter's house. But after he had entered and was seated, he became apprehensive of Hsi's objections, and kept it in his bosom, not daring to bring it out. After he was outside the door he threw it back from a distance, then turned around and walked hastily away." (Richard B. Mather, "Chinese Letters and Scholarship in the Third and Fourth Centuries: *The Wen-Hsüeh P'ien* of the *Shih-Shuo-Hsin-Yü*." *Journal of the American Oriental Society*, IV (1964), 353.) But another version of the story is recorded in *Wei Chih*, a history of the Wei dynasty: "Chung Hui was an intimate friend of the Generalissimo (Ssu-Ma Chao). When he heard of Hsi K'ang's fame, he went to visit him... He arrived in a magnificent chariot and in elegant dress, followed by a cloud of friends. Hsi K'ang was just sitting on the floor with his legs spread out, engaged in forging iron, and when Chung Hui arrived, he did not pay him any respect. (When after waiting in vain Chung Hui turned to go), Hsi K'ang asked him: "What did you hear that you came, and what did you see that you go away? ", whereupon Chung Hui answered: "Hearing what I did I came, seeing what I see I go." (Van Gulik, 17.)

15. Chung Hui's role in this is recorded in the *Shih Shuo Hsin Yü* and translated by Van Gulik, p. 19: "Then Chung Hui impeached Hsi K'ang before the Generalissimo, saying: 'Now the Imperial sway is unfolded brilliantly, a soft breeze blows over the four seas. In distant regions there are no wily and obsequious people, in the streets of the capital no dissenting voices are heard. But Hsi K'ang above does not loyally serve the Emperor, and below does not serve kings and princes. He is overbearing, thinks himself too good for this world, and does not work for the common good... If now Hsi K'ang is not executed, there is no possibility of making clear the Kingly Way.' "

16. Hightower, 162.

17. This is included in the WH, chuan 23 and translated into German by E. Von Zach, Sinologische Beitrage, II, (Batavia, 1935) and reprinted by Harvard University Press as No. XVIII of the Harvard Yenching Institute Studies as *Die Chinesische Anthologie; Ubersetzungen aus dem Wen hsüan* (Cambridge Mass., 1958) 2 vols.

18. Van Gulik, 22.

19. Van Gulik, 24.

20. Richard Wilhelm, *The I Ching or Book of Changes*, Cary F. Baynes trans. (Princeton, 1967), 17:

Clouds and Thunder:
The Image of *Difficulty at the Beginning*
Thus the superior man
Brings order out of confusion.

The character *lun*, "to discourse" is etymologically related to *lun*, to set in order. The original meaning was to twist or to wind silken threads, from that it came to mean to classify or to put in order, and from that to discourse. The "word" radical was added to the written form for the latter.

21. E.R. Hughes, *The Art of Letters Lu Chi's "Wen Fu," A.D. 302; A Translation and Comparative Study* (New York, 1951), 232.

22. Vincent Yu-chung Shih trans., *The Literary Mind and the Carving of Dragons by Lii Hsieh: A Study of Thought and Pattern in Chinese Literature* (New York, 1959), 101.

23. Shih, 102-103.

24. *Meng Tzu*, Book I, Chapter 3, verse 3.

25. James Legge, *The Texts of Taoism* (New York: Dover reprint, 1962), I, 198-202 translates the chapter title as "Nourishing the Lord of Life." He explains the "Lord of Life" as Tao (p. 130). I prefer to translate *chu* as basis or foundation, following a passage in the Appended Judgments, Chapter 8 verse 5 of the *Book of Changes:* "Words and actions are the hinge and spring of the superior man. The movement of that hinge and spring is the basis of glory or disgrace." Thus the chapter title in the *Chuang-tzu* should be "The basis of the Nourishment of Life."

26. Legge, I, 198.

27. Legge, I, 200. The third section contains a parable of a wild pheasant: "A pheasant of the marshes has to take ten steps to pick up a mouthful of food, and thirty steps to get a drink, but it does not seek to be nourished in a coop." Legge, I, 200. Thus *Yang-sheng* means the nourishment of a free life. The fourth section contains the story of Lao-tzu's death. Here Legge's translation is superb: "When the Master came it was at the proper time; when he went away, it was the simple sequence (of his coming). Quiet acquiescence in what happens at its proper time, and quietly submitting (to its ceasing) afford no occasion for grief or joy." Legge, I, 201. This introduces into Yang-sheng the notion of a "proper time" which is an *I Ching*, or *Book of Changes* concept. See Hellmut Wilhelm, "The Concept of Time in the Book of Changes," *Man and Times, Papers from the Eranos Yearbook* (New York, 1957), 212-232. On the nourishment aspect of *Yang-sheng* see Henri Maspero's epoch making article: Les procédés de "nourir le principe vital" dans la religion taoïste ancienne *Journal Asiatique* CCXXIX (1937), 177-252. On the treatment of Taoism in general see Joseph Needham and Wang Ling, *Science and Civilization in China*, Vol. 2 (Cambridge, 1962).

28. *Ssu Pu Ts'ung K'an* ed. (hereafter SPTK) 7.10b.

29. Edward H. Schafer, *The Vermilion Bird; T'ang Images of the South* (Berkeley, 1967), 7 discusses the notion of *hsien.* "Transcendent" is one of the most noteworthy of them. This represents the word *hsien*, which others have commonly rendered "(Taoist) immortal." The word actually connotes "having the ability to spring up out of the mire of the material world into the realm of spirit, symbolized by unearthly journeys to sky castles and island paradises," with linguistic cognates meaning "soaring, flying, flapping up." In Han times it had ordinarily been used as a verb to characterize the activity of the "Feathered Men," who were Taoist adepts and initiates, transformed men shown pictorially with angelic wings. I shall normally translate the word by "transcendent" (an equivalence which I did not invent, I am sorry to say), though in a few places I have used "sylph" to suggest the dainty, refined inhabitants of the upper air. Because of its use in the Western philosophical tradition I have not followed Schafer in rendering *hsien* as transcendent, but have stayed with *immortal* which seems to be a more neutral word. On the feathered *hsien* as a motif in Han art see: Florence Waterbury, *Bird-Deities in China* (Ascona, Switzerland, 1952).

30. Legge, I, 171.

31. Yü Ying-shih, "Life and Immortality In the Mind of Han China." *Harvard Journal of Asian Studies,* XXV (1964-65), 95-96.

32. Yü Ying-shih, 102-03.

33. Yü Ying-shih, 104.

34. See P. Van der Loon, "On the Transmission of the Kuan Tzu," *T'oung Pao*, XLI (1952), 357-393, at 368n.

35. *Sui Shu, (Jen shou* ed.) 34.28a-34a. The two legendary rulers Huang Ti and Shen Nung were both associated with agriculture. Shen Nung was the reputed founder of agriculture, but Huang Ti, who came later, improved upon his work by determining the proper time for grain and tree planting. Shen Nung was the first to teach the people the medicinal value of certain plants. See E.T.C. Werner, *A Dictionary of Chinese Mythology* (Shanghai, 1932), 419.

36. A difficulty arises however in that the two works quoted by Li Shan do not occur in the Bibliographical Treatise of the Sui dynastic history (Sui Shu). This treatise was compiled on an imperial order of A.D. 641 by Chang-sun wu-chi (d.659) and others and presented to the throne in 659. Li Shan wrote his commentary sometime between 656 and 660. Thus we should expect books known to Li Shan to be extant in the imperial library. And so the problem is: how did Li Shan quote from the *Nourishment of Life Classic* and the *Essentials of the Nourishment of Life*? We know from independent sources of the existence of the *Nourishment of Life Classic* (from a note to the text of the *Po wu chih*, a collection of anecdotes) so it was not fabricated by Li Shan to provide embellishment for his commentary. Probably the easiest solution to this problem is that Li Shan himself was quoting material that was lost but was recorded in an encyclopedia of quotations or in a collection of quotes from a specific work. In the case of the *Yang-sheng yao*, or *Essentials of the Nourishment of Life*, there is a work called the *Yang-sheng yao chi* which might be translated *Collections From The Yang-sheng yao* by a certain Chang shen of the latter Han dynasty. This work *is* mentioned in the Bibliographical Treatise, but is now lost. Although there is a difficulty in pinpointing exact texts, the conclusion remains that there was a well defined Yang-sheng literature of which Hsi K'ang was well aware.

37. The only available translation of the *Yang-sheng lun* is by Donald Holzman, *La Vie et la Pensée de Hi K'ang* (Leiden, 1957), 83-91. Holzman also translates Hsiang Hsiu's critique and Hsi K'ang's further response. Wing-tsit Ch'an has translated several paragraphs of Hsiang Hsiu's critique in Wm. Theodore de Bary ed. *Sources of Chinese Tradition* (New York, 1960), I, 248-49. Aside from the WH, the *Yang-sheng lun* appears in the *Ku-chin-Wen-Hsüan*, or the *Ancient and Modern Literary Anthology*, Liang Jung-jo ed. (Tai ei, 1962), IV, 1831-1838, and *Lu Hsün Chüan Chi*, or *The Collected Works of Lu Hsün* (Shanghai, 1946), IX, 53-57. It also appears in several other collections. This translation attempts to preserve, as much as possible, the concreteness of the Chinese literary images. To be sure the result is not always idiomatic English, but it attempts to reflect, as nearly as the translator is able, the literary craftsmanship and philosophical standpoint of Hsi K'ang as well as the semantic, morphological, and syntactical structures of the Chinese. Because this essay was selected by Prince Hsiao T'ung for his WH we may presuppose its literary excellence; there are however literary allusions the meanings of which have been lost and I have not attempted to throw a logical net around this essay in an attempt to give explanations where none can be given.

38. A work called the *T'ien yin tzu* attributed to an unknown author of T'ang times attempts to distinguish between five different kinds of *hsien:* "Among men they are known as *hsien.* In the heavens they are known as *'heavenly hsien.'* On the earth they are known as *'earthly hsien.'* In the waters they are known as *'water hsien.'* Those who are able to universally change are known as *shen hsien*, or 'spiritual immortals' ". *(P'ei-wen yün-fu*, or *A Riming Book of Literary Allusions* Taiwan reprint of the Commercial Press ed., I, 697.2.) I have used the term immortal throughout this translation since the above is a much later work and we cannot be sure this interesting set of distinctions were in actual use during Hsi K'ang's time.

39. Yü Ying-shih, p. 90 states the term *Pu-ssu* was a technical term of late Chou and Han Taoism. It occurs as a technical term in the *Han-fei tzu*, or *Book of Master Han Fei*, as a "no death drug." In the *Shan-hai ching*, or *Classic of Mountains and Seas*, as "no death people," "no death country," "no death mountain," "no death tree," and "no death water." It occurs in the *Huai-nan tzu* as "no death grass."

40. Li Shan here quotes from the *Nourishment of Life Classic:* Huang-ti in questioning T'ien-lao (his chief minister) said: "In human life the eldest longevity is 120 years; the middle longevity is one hundred years; the youngest longevity is eighty years and those who are not like this all die

young." Later Li Shan quotes from another section of the *Nourishment of Life Classic* (see note 44) which seems to contradict this statement. Thus we see that Hsi K'ang while following the yang-sheng literature, was not totally committed to *hsien* Taoism. As we shall see later, he was very close to the Chuang tzu school of "relativity of knowledge."

41. The History of the Former Han Dynasty *(Han Shu)* in the monograph on the suburban sacrifices says: "In the talk of the Taoist magicians *(fang-shih)* of the states of Yen and Ch'i there are the *shen hsien*, or immortals." *Han Shu, Po-na* ed. 25a, 19b.

42. *I ch'i* is given a literal translation here. *I* is often used in Chinese literature in apposition with *tsai* where *i* and *tsai* are both used to name calamities from heaven. *I* means events which are extraordinary but which bring no direct loss of life. Such things as eclipses, two-headed animals, etc. are classed as *i*. *Tsai* are also extraordinary events which do bring loss of life, i.e. flood, famine, etc. On *ch'i*, Wing-tsit Chan trans., *Reflections on Things at Hand; The Neo-Confucian Anthology* (New York, 1967), 360 says: "In many cases, especially before the development of the Neo-Confucian doctrine of *li*, *ch'i* denotes the psychophysical power associated with blood and breath." The term *i-ch'i* also appears in the *Pao-p'u tzu*, or *The Master of Maintaining Simplicity (Ssu-Pu Ts'ung-K'an* ed., - hereafter SPTK) 3.lb line 5. Although the *Pao-p'u tzu* was written about A.D. 320, it is still relevant for semantic comparison.

43. It is possible that *Tao-yang* refers to *Tao-yin yang-sheng*, or "calisthenics and the nourishment of life." *Tao-yang* appears twice in the *Pao-p'u tzu* (SPTK ed.), 13.3b line five and 3.5a line nine, and is translated by James Ware, *Alchemy, Medicine, Religion in the China of A.D. 320: The Nei P'ien of Ko Hung (Pao-p'u tzu)* (Cambridge, Mass., 1966), 216 and 59, as "calisthenics and diet." Eugen Feifel, "Pao p'u tzu, nei p'ien, Chapter I-III," *Monumenta Serica*, VI (1941), 193, translates Tao-yang as "breathing and nourishing the life fluid."

44. Here Li Shan quotes another passage from the *Nourishment of Life Classic* which seems to contradict the previous one: "The largest period of a human lifespan has as its limit one hundred twenty years. If one economizes and protects it, one can reach a thousand years." These differing points of view lead one to suspect that Li Shan is quoting from fragments and further, that some of these fragments may not be authentic.

45. Li Shan quotes a story from the *Shih chi*, or *Historical Records*. Burton Watson trans., *Records of the Grand Historian of China; the Shih Chi of Ssu-Ma Ch'ien* (New York, 1961), I, 165-66: "Again Chou P'o apologetically replied, 'I do not know,' and the sweat poured down and drenched his back, so chagrined was he at being unable to answer." This event probably has nothing at all to do with Hsi K'ang's meaning, but is used here by Li Shan to give a gloss to the words. The fang-shih had drugs for producing sweat, which were symbolic of a drought ending rain. See note 75.

46. This appears in the *Li chi*, or *Records of the Rites*. James Legge trans., *The Sacred Books of the East, III, The Li Ki* (Delhi, India reprint, 1966) I, 133: "Tseng tzu said to Tzu ssu, 'Khi, when I was engaged in the mourning for my parents, no water or other liquid entered my mouth for seven days.' "

47. An earlier occurrence of this phrase occurs in the *Shih chi*, or *Historical Records (Chung-hua shu-chu ed.)* 81.2440. Here Lin Hsiang-ju has taken a precious jade to the king. The king takes the jade, intending to keep it as a gift. Lin tells the king it has a flaw and asks permission to show it to him. "The king gives him the jade. Hsiang-ju uses the opportunity to seize the jade and step back. He leans against the pillar and is so angry his hair pushes up his cap." Lin then threatens to break the jade if the king does not pay for it by ceding territory to his state of Chao.

48. The reign of the mythological emperor Yü was witnessed by ten years of floods and the reign of the emperor T'ang by seven years of drought. Yü was sent out to drain the floods by the mythological Emperor Shun which he accomplished only after nine years of hard work. Emperor T'ang was the

founder of the Shang dynasty, traditional dates 1766-1122 B.C.

49. Li Shan cites a comparable passage in the *Chuang tzu.* James Legge, *The Texts of Taoism,* (Dover reprint, 1962) I, 364: "Blowing and breathing with open mouth; inhaling and exhaling the breath; expelling the old breath and taking in new; passing their time like the (dormant) bear, [lit. bear passing, bird stretching; these might have been names for specific breathing exercises] and stretching and twisting (the neck) like a bird;—all this simply shows the desire for longevity. This is what the scholars who manipulate their breath, and the men who nourish the body and wish to live as long as Peng Tzu are fond of."

50. Li Shan quotes a stanza from an ancient poem the title of which is not cited: "dietary discipline seeks after immortality."

51. A measure for corn normally holding from five to ten pecks depending upon the times.

52. Hsi K'ang is quoting a book by Fan Sheng-chih commonly called *Fan Sheng-chig shu,* or *Mr. Fan Sheng-chih's Book.* The existing fragments of this book have been brought together and translated by Shih Sheng-han, *On 'Fan Sheng-chih Shu' An Agriculturistic Book of China Written By Fan Sheng-chih in the First Century B.C.* (Peking, 1959), 29-31. 7.1 "During the reign of Emperor T'ang, there was a long spell of severe drought. (The Prime Minister) Yiyin developed the system of *'ou-t'ien,'* [or ch'ü field], i.e. cultivation in shallow pits, taught the people to treat the seeds and to carry water for irrigating the crop.' (sic.) 7.1.8 With this system of cultivation in shallow pits, one must always see to watering (sic.) of his crop in days of drought. A yield of 100 *hu* per *mou* [acre] may be expected (if diligently worked for)."

In his analysis of the subject matter in the *Fan Sheng-chih shu,* Professor Shih, p. 64, says: "The original meaning of the word *'ou'* [or *Ch'ü*] is a depression or a shallow pit made to receive or detain something. The significance of such a pit in retention of water and nutrients for plant growth is self-evident... In certain localities, for example at headwaters, or mountainous districts with little level ground and heavy erosion, and where man power is ample while resources of manure are scanty, this sort of intensive husbandry may be considered as a means to directly increase the income of local farming people and indirectly to reduce the threat downstreams of flooding induced by erosion. But the intense toil in which no animal service or machine work is possible, cannot be welcome if it is to produce the food cereal of no great market prices. (sic.) Therefore, the method of cultivation in shallow pits ought not to be looked upon as the singular effective way to produce a high yield. Unless well-contemplated in connection with other conditions, any hasty recommendation might turn out to be very harmful." Exactly what Hsi K'ang had in mind with this allusion is not clear to me. Presumably this "*ch'u* field" method would have been known to farmers of his time if it was in fact effective. Perhaps it had been lost as a method by Hsi K'ang's time and was recorded only in this book. Of course his point is the significance of the differing methods of nurturing the seedlings, an instance of the Chuang tzuian "relativity of knowledge." The reference to sowing grain in the time of T'ang mentioned in paragraph four of the translation is probably connected with the above quoted passage.

53. The *Chung Kuo yao hsueh ta tzu tien,* or *Chinese Materia Medica* (Shanghai, 1935), 1452 says the elm tonic causes one to go to sleep. Used daily it is a cure for an inflamed gall bladder. On beans causing heaviness, Li Shan quotes the *Po wu chih,* or *The Record of the Investigation of Things:* "If one eats beans for three years then the body will become heavy and starting and stopping will be difficult."

54. *Acacia Nemu.* E. Bretschneider, *Botanicon Sinicum* (London, 1882), III, 480: "This tree grows in Yü chou (Honan province) in mountain valleys. It resembles the *kou-ku* (which is an *Ilex).* The bark is used in medicine. Taste sweet, nature uniform, non-poisonous. The leaves and flowers are likewise officinal." It is also identified with *Albizzia Julibrissin* and *Acacia Julibrissin.* See *Pen-tsao kang-mu,* or the *Great Pharmacopoeia* (Hu-pao chai, 1893 ed.) 35B. 1b, (hereafter PTKM).

55. *Hemerocallis flava, L.* This is the common lily and was once called "The Forgetter of Sorrows." Li Shan quotes a work called the *Shen Nung Pen Ts'ao* (see note 60) which says: "The *Ho-huan* removes anger, the *Hsüan* herb causes one to forget sorrow." For *Hemerocallis* as a motif in later art see Bo Gyllensvärd, "A Botanical Excursion In the Kempe Collection," *Bulletin of the Museum of Far Eastern Antiquities*, XXXVII (1965), 183. The lily is now known to have originated in China and later spread to the Mediterranean. It appears in the Book of Odes, but under another character with the same pronunciation.

56. *Spheroides vermicularis.* It is, I am told, used in a soup, especially in Southern China. On "strong leeks" or *Hun hsin lei*, Bretschneider, I, 59 lists thirty eight species.

57. Because it is constantly turning around to look for danger the neck develops an irritation.

58. Because the people of Chin were fond of eating a certain kind of date, their teeth became black. These last three allusions may possibly refer to proverbs current among the people of the time but now lost.

59. Besides being the inventor of the plough and of markets, Shen Nung, or the Divine Agriculturalist discovered the five kinds of grain and developed the use of herbs for medicinal purposes. Hsi K'ang here extols Shen Nung for his knowledge of the medicinal aspects of certain plants and herbs, while Shen Nung's development of the five grains is antithetical to Taoist interests. See note 60.

60. This quotation comes from the *Shen Nung pen ts'ao ching*, or *Classical Work of Medicines by the Emperor Shen Nung.* This was probably a Han time work attributed to Shen Nung (in order to derive legitimacy from his legend.) Emperor P'ing ti in A.D. 4 ordered all men familiar with medical prescriptions and the oral traditions of the *pen ts'ao* to assemble at the capital. It may well have been written at that time. The PTKM collects all the surviving fragments of the Shen Nung work and the fragment quoted by Hsi K'ang in part has been translated by Bretschneider, I, 29-30 in full: "Of the first class of drugs there are 120 sorts which are considered to perform the functions of *chün* or Sovereigns. They support human life, and thereby resemble Heaven. They are not poisonous. Whatever quantity you take, or howsoever long you use them, they are harmless. If you wish to have the body light, to improve the breath, to live to an old age, without growing old, make use of these drugs of the first class. There are 120 sorts of drugs of the second class, which are considered to perform the functions of *ch'en* or Ministers. They support human nature, and thereby resemble man. Some of these drugs are harmless; others are poisonous. They ought to be used with discretion. If you wish to lessen the violence of diseases and re-establish decayed strength, use the second class of drugs. There are 125 sorts of drugs of the third class which are considered to perform the functions of *tso,* Assistants and *shi,* agents. They cure diseases, and thereby resemble earth. They are very poisonous and ought not to be used continuously. If you wish to drive out cold or heat from the body, or to correct the breath, or to open obstructions, or to heal diseases, use the drugs of this class."

61. Holzman, p. 87, has a slightly different rendering of this passage: "he truly understood the natural order of human nature and destiny; and it is in helping (rational) nourishment that he succeeded." Holzman thus construes *t'ung* as "to succeed." The term *fu-yang* seems to be used in a technical sense here, but I can find no further relevant sources for this.

62. Lists of the five grains vary. Probably they were hemp, millet, wheat, barley and beans. For the addiction to sounds and colors Li Shan cites out of context a phrase in the *Chuang tzu;* Legge, II, 182: "Moreover, music, beauty, with the pleasures of the taste and of power, are appreciated by men's minds and rejoiced in without any previous learning of them."

63. *Tzu-jan,* or nature (or spontaneity, depending upon the context) is a crucial concept in Taoism. Literally "self-sameness", its *locus classicus* is the *Lao Tzu*; Legge, I, 68: "Man takes his law from the Earth; the Earth takes its

law from Heaven; Heaven takes its law from the Tao. The law of the Tao is its being what it is *[Tzu-jan]*. See also Joseph Needham, *Science and Civilization in Ancient China* (Cambridge, 1956), II, 51 *passim.*

64. Duke Huan's dates are 684-642 B.C. He was a Duke of the state of Ch'i. Pien Ch'üeh's real name was Ch'in Yüeh-jen. According to tradition he was originally an innkeeper who received from one of his customers some old books on medicine as well as some fabulous drugs. Travelling from state to state he became famous as a doctor. He was supposed to have been able to see within the body of his patients to locate the disease and determine which direction it was moving.

65. This story appears in the *Han-fei tzu* and is translated by W.K. Liao, *Han Fei Tzu: Works from the Chinese* (London, 1959), I, 214-215: "Pien Ch'iao *[Pien Ch'üeh]* once had an interview with Duke Huan of Ch'i. After standing for a while, Pien Ch'iao said: 'Your Highness has a disease in the capillary tubes. If not treated now, it might go deep.' 'I have no disease,' replied Duke Huan. After Pien Ch'iao went out, Duke Huan remarked: 'Physicians are fond of treating healthy men so as to display their attainments.' Ten days later, Pien Ch'iao again had an interview and said: 'The disease of Your Highness is in the flesh and skin. If not treated now, it will go still deeper.' To this advice Duke Huan made no reply. Pien Ch'iao went out. Duke Huan was again displeased. After ten more days, Pien Ch'iao had another interview and said: 'The disease of your Highness is in the stomach and intestines. If not treated now, it will go still deeper.' Again Duke Huan made no reply to the advice. Pien Ch'iao went out. Duke Huan was again displeased. After ten more days, Pien Ch'iao, looking at Duke Huan, turned back and ran away. The Duke sent men out to ask him. 'Diseases that are in the capillary tubes,' said Pien Ch'iao, 'can be reached by hot water or flat irons. Those in the flesh and skin can be reached by metal or stone needles. Those in the stomach and intestines can be reached by well-boiled drugs. But after they penetrate the bones and marrow, the patients are at the mercy of the Commissioner of Life (the name of a star that was supposed to superintend the life and death problem of every mortal) wherefore nothing can be done. Now that the disease of His Highness is in his bones and marrow, thy servant has no more advice to give.' In the course of five more days, Duke Huan began to feel pain in his body, and so sent men out to look for Pien Ch'iao, who, however, had already gone to the Ch'in State. Thus ended the life of Duke Huan."

66. Hsi K'ang is here most probably referring to the whole corpus of the *Yang-sheng* tradition rather than merely the nourishment of life as a non-technical term.

67. Li Shan explains that *Wei lü* was a fiery rock in the sea which evaporated all the excess water collected by countless rivers pouring into it. This occurs in *Chuang tzu*, Legge, I, 375: "A myriad streams flow into it without ceasing, and yet it is not filled; and afterwards it discharges them (also without ceasing), and yet is not emptied." The same image is used in the *Pao-p'u tzu* (SPTK ed.) 2.8z; Ware, p. 43: "But gills of aid do not meet barrels of expense, just as an irrigation ditch could never carry away the water that leaks out of the sea at Wei-lü and prevents the sea from overflowing." Ko Hung might well have taken this particular usage from Hsi K'ang rather than from the *Chuang tzu* since Hsi K'ang's meaning is identical with his own. *Wei-lü* is also explained as a large drain hole (a "world navel") through which excess waters flow.

68. The *Yü chang* tree was often confused with the *Machilus nanmu*, a species of cedar. The *Catalpa Kaempferi*, associated with the *Yü chang*, is no longer called by that name. The tree occurs in the *Chuang tzu*; Legge II, 36: "Has your majesty not seen the climbing monkey? When he is among the plane trees, rottleras, oaks, and camphor trees, he grasps and twists their branches (into a screen), where he reigns quite at ease..."

69. *Yu yu che* means those with fixed aims. *Yu* by itself is a common term in the *Book of Changes* having the meaning of one's aim, goal, or destination.

70. These terms are all from the *Chuang tzu*; Legge, I, 298: "You must be still; you must be pure; not subjecting your body to toil, not agitating your vital force; - then you may live for long."

71. That is, he nourishes himself with the harmony which comes from the proper relation of knowledge to placidity. Li Shan cites the *Chuang tzu*; Legge, I, 368-69: "The ancients who regulated the Tao nourished their faculty of knowledge by their placidity, and all through life abstained from employing that faculty in action; —they must be pronounced to have (thus also) nourished their placidity by their knowledge. When the faculty of knowledge and the placidity (thus) blend together, and they nourish each other, then from the nature there come forth (sic.) harmony and orderly method. The attributes (of the Tao) constitute the harmony; the Tao (itself) secures the orderly method.

72. This term occurs in the *Lao tzu.* Legge, I, 109 translates it as "great conformity": "Deep and far reaching is such mysterious excellence, showing indeed its possessor as opposite to others, but leading them to a great conformity to him."

73. *Gyrophora vellea, Ach.* This plant is a moss or lichen, although sometimes called a mushroom in popular literature, and grows in the T'ien T'ai mountain range. According to tradition it grows only on granite cliffs and is called "stone's ear", or *shih er* because it hangs like an earring. The *ling-chih* plant in later times is associated with an art motif; Bo Gyllensvärd, 187: "An ornament known to the Chinese as *ju-i* is often found in different contexts. It forms the principle motif of a sceptre and as such expresses the attainment of wishes, e.g. for long life. Many attribute its shape to a fungus, Ling-Chih, which is also often used as a symbol of happiness." Pan Ku, author of the *Han Shu* or *History of the Han Dynasty* wrote a poem in praise of *Ling-chih* (which is preserved in the *P'ei-wen yun-fu*, I, 116, and in the *Ch'uan han san kuo chin nan-pei ch'iao shih*, or *The Complete Poems of the Han, San Kuo Period, Chin and North-South Dynasties,* Ting Fu-pao comp. (Wu hsi, 1919), 2.8b-9a. A tentative translation of this might read: "Because of a heavy dew, *Ling-chih* is produced. Symbol of the three powers. An auspicious omen. Prolonging one's life and illumining this capital, it matches the Lord on High and images the Great Infinity. A trio with the sun and moon, it diffuses its brightness." (The three powers were the power of heaven, earth, and man.) *Ling-chih* also seems to have been a generic name for such species as the "Blue Cloud *chih*", the "Dragon Fairy *chih*", the "Gold Orchid *chih*", the "Spirit of the Moon *chih*", the "Night Light *chih*" which had luminous qualities (possibly this is what Pan Ku is referring to in the second stanza), the "Fire Fly *chih*", the "White Cloud *chih*", and several others. For an interesting summary of the popular lore of the *Ling-chih,* see Alfred Koehn, *Fragrance From A Chinese Garden* (Peking, 1942), 52-54. Whether Hsi K'ang is here referring to the *Ling-chih* of popular lore or the *Gyrophora vellea* is difficult to say. By steaming the *Ling-chih* one could, presumably, obtain Gyrophoric acid, but what this would do to a person is unknown to me.

74. What is meant here seems to be a good wine or a strong wine.

75. Li Shan says this phrase means to "nourish harmony among the creatures" which makes no sense. From the context it appears that this term as well as "steaming oneself with the *Ling-chih* plant" and "drinking the unfermented wine" are names, possibly of Taoist exercises. Ritual exposure in earlier religion was designed to end drought by exposing oneself to the morning dew and the morning sun. That is, by standing naked in the heat of the day, the shamans hoped for mercy from heaven in the form of rain. When this proved unsuccessful the shaman resorted to self-immolation. This is probably the source of a specifically Eastern form of political rejection. See Needham, II, 144-45 and Edward H. Schafer, "Ritual Exposure in Ancient China," *Harvard Journal of Asiatic Studies*, XIV (1951), 130.

76. *Wu-wei* and *Tzu-te* are two important concepts of Taoism. Li Shan quotes from the *Chuang tzu* to illustrate the use of *Wu-wei,* to do nothing or

non-interference; Legge II, 3: "Heaven does nothing, and thence comes its serenity; Earth does nothing, and hence comes its rest. By the union of these two inactivities, all things are produced." *Tzu-te,* or self-contentment, occurs ten times in the *Chuang tzu* as opposed to four times in the *Meng-tzu.* The emphasis on *Tzu-te* in *Chuang tzu* may be illustrated with the following quote from Legge's translation, I, 291: "The government of the world by Kieh, (on the contrary), made men look distressed; but when their nature shows the symptoms of distress, there is a want of its (proper) contentment. The want of placidity and the want of contentment are contrary to the character (of the nature); and where this obtains, it is impossible that any man or state should anywhere abide long."

77. *Hsüan,* or mysterious, dark, is another important technical term of Taoist provenience. Li Shan quotes its *locus classicus* in the *Lao tzu*; Legge, I, 47: "Where the mystery is the deepest is the gate of all that is subtle and wonderful."

78. The origin of this term is in doubt. Needham, II, 133 suggests it might be an early transliteration of the word shaman in Chinese. In any case Hsi K'ang's time regarded him as a historical figure that had attained the state of *hsien.*

79. Wang Tzu-ch'iao was a famous immortal whose biography is preserved in the *Lieh hsien chuan,* or *Biographies of Illustrious Immortals (Ts'ung Shu Chi Ch'eng* ed.). A. 23. The historical Wang Tzu-ch'iao was the eldest son of King Ling of the Chou dynasty (6th. century B.C.) He is said to have studied for thirty years under a magician named Fou-ch'iu Kung. See M. Kaltenmark, *Le "Lie Sien Tchuan" ['Lieh Hsien Chuan']; Biographies Légendaires des Immortels Taoistes de l'Antiquité* (Peking, 1953), 109.

University of Washington,
Seattle, Washington.

Will's Criticisms of Peirce's Frequency Theory of Probability

by Richard Tursman

● *The purpose of this essay* ● The purpose of this essay is to defend Peirce's frequency theory of probability against two criticisms which have been brought against it by Professor F. Will; Will's criticisms were published in *The Journal of Philosophy* (volume LXII, no. 3, 1965)

● *Peirce's frequency theory of probability* ● Peirce's theory of probability is really a theory about two different, though intimately related, kinds of probability: one of these two kinds is probability in a technical sense, or objective probability; the other kind is probability in an untechnical sense, or probability in the sense of reasonability. The general claim that Peirce holds a frequency theory of probability is true by virtue of the fact that he holds a frequency theory of both of these kinds of probability and that in various subtle ways the frequencies involved in each of the two kinds of probability are related.

It is real (in the sense of Peirce's scholastic realism) that if water is heated a certain way under certain conditions, then it will boil. Objective probability is real in the same sense, because it too is the fact that if a thing is done a certain way under certain conditions, then a certain result will follow. Peirce puts it this way, "if of an endless series of possible experiences a definite proportion will present a certain character (which is the sort of fact called an objective probability)...." (Collected Papers 2.785 (2)) Roughly speaking, the thing to be done is to sample a population in an unbiased way and under the conditions enjoyed by an unhampered scientist. The sort of sample desired is one chosen such that every other possible sample of that size or description was just as likely to have been chosen. If the selection of a sample is really unbiased, then we have the desired sort of sample, namely, a random sample. The person doing the sampling pre-designates the population he will sample and the characteristics he will be looking for in such a way that he is able, as he finishes his sampling, to record that such and such a proportion of the cases examined had the predesignated characteristics. One of Peirce's own examples pre-designates as the population to be sampled a cargo of wheat which has been stirred with great thoroughness and pre-designates the characteristic to be looked for that of being quality A wheat (or not). The result, in the case of Peirce's example, was that of a sample of twenty-seven thimblefuls of wheat taken from evenly distributed sections of the cargo 4/5 was of quality A.

The clearest idea we have of objective probability and the best definition we can give of it center around the consideration of the result of a particular action. We consider that if we were to continue indefinitely to sample this cargo of wheat, or some analogous population, so that in effect the number of our samples increases indefinitely (we throw our samples back and re-mix every time!), then the values attained when plotted will form a normal distribution curve around the actual population average. The actual population average is defined as the limit of the indefinitely long series of ratios attained by sampling. Knowing that we would, in the long run, come up with this figure, we define the objective probability of the wheat's being quality A as the limit. Since we never really reach this limit in practice, we estimate the population average using the sample average so far obtained as our estimator. Even if we had randomly selected our samples we would still need a legitimate mathematical way to move from our data as obtained from our sampling up to a certain point to the population average. Fortunately the sample average is, mathematically, an unbiased estimator and is, in fact, all else being equal, the best estimator unless we have some additional information on the basis of which we could make a range adjustment. Objective probability, then, is generally defined as the relative frequency with which a certain characteristic would occur in the long run under conditions such as those just outlined. Because Peirce thus defines objective probability in terms of frequency, it is clear that he holds a frequency theory of *objective* probability.

Peirce also holds a frequency theory of untechnical probability, or probability in the sense of reasonability. Of untechnical probability he gives the following account:

> In the propriety of language, the probability of a fact either is, or solely depends upon, the strength of the argument in its favor, supposing all relevant relations of all known facts to constitute that argument. Now, the strength of an argument is only the frequency with which *such* an argument will yield a true conclusion when its premises are true. Hence probability depends solely upon the relative frequency of a specific event (namely, that a certain kind of argument yields a true conclusion from true premises) to a generic event (namely, that that kind of argument occurs with true premises). Thus, when an ordinary man says that it is highly probable that it will rain, he has reference to certain indications of rain —that is, to a certain kind of argument that it will rain— and means to say that there is an argument that it will rain, which is of a kind of which but a small proportion fail. "Probability," in the untechnical sense, is therefore a vague word, inasmuch as it does not indicate what one, of the numerous subordinated and coordinated genera to which every argument belongs, is the one the relative frequency of the truth of which is expressed. (3.19)

Peirce is looking back down the road from the mathematical continuum of technical probability toward non-quantitative logic and beyond to the realm of *logica utens.* Where he stands, things are clear. He is able to form a picture of scientists taking random samples of populations, discovering that a proportion r of them have certain characteristics, estimating that probably and approximately r of the population has the same characteristics, continuing this process, and correcting past results such that in the long run the population average would be reached. As the sample becomes less than numerous, the relative of the induction is not longer a probable statistical deduction but merely a probable deduction. When quantities are no longer known, inferences are based upon our experiences of "some", "most", "almost all", and similar sorts of samples. The controls go; scientific procedures are replaced by inferior methods of fixing belief; alchemy substituted for modern methods of chemical fixation. Leading principles are lost sight of and eventually the only thing left is habit.

Now probability in the sense of reasonability would be given a place on this road slightly above habit. At this place we must imagine the amateur inquirer doing such things as wondering whether or not it will rain and presenting, perhaps, some simple arguments pro or con. He might remember cases in which under the same general conditions, such as clouds, thunder, etc., he had believed that it would rain together with the result in each case. He might recognize similarities between his arguments and other arguments of vaguely the same type. The following passage, in agreement with the above, makes it clear that Peirce would *associate* such untechnical reasoning with technical probability, the main reason being that frequency is involved in both cases.

> If somebody replies that in weighing arguments *pro* and *con* critics make no use of the mathematical calculus of probabilities, the rejoinder will be that their proceeding only differs from that by its greater vagueness, and that a vague and inexact use of probabilities has no logical advantage over a more critical employment of them. (7.181)

We have made the point that frequency is involved in a vague way, but we now assert the really crucial point that the series wherein we examine frequency need be no longer than the case in question plus one other case. In other words, probability in the sense of reasonability requires minimally what is in effect nothing more than a comparison of the case in question with a single analogue in order to insure that frequency is involved. In fact, all Peirce denies is that there can be reasonability without the relating of the case in question to at least one previous case. (He maintains the same thing, I think, in regard to both technical and untechnical probability, the difference being a matter of vagueness in the case of untechnical probability.)

So far, we have been describing Peirce's view of objective probability and his view of untechnical probability; we have indicated that frequency is at the bottom of both kinds of probability. An alternative way of describing Peirce's view of probability in general would have been to have spoken of technical and untechnical probability as two aspects, one clear, one vague, of the same kind of probability. If we had adopted this alternative description, then we could have constructed one argument for each grain of wheat, one argument to the effect that that grain of wheat was of quality A, and then it would have turned out that our arguments would have been correct 4/5 of the time; and we would still have estimated the population average to have been 4/5. This alternative description has the advantage of showing various close connections between technical and untechnical probability, especially at those points at which the two merge into one another. But this alternative description would not illustrate as well the difference between technical and untechnical probability.

Because of the deductive certainty of the central limit theorem; and because of the deductive certainty of the functioning of the sample average as an estimator of the limit; the difficulty in establishing the reasonability of this frequency theory of objective probability centers around the difficulty in establishing (1) that any given sample is (and will be) a random sample of the population in question, and (2) that it is reasonable to follow an inference which by its very nature refers to a future which we shall never reach.

Granted that the inquirer never knows for sure that his sample *is* random, Peirce tries to defend the reasonability of the procedure on the grounds that the *scientific* inquirer tries as best he can to obtain unbiased samples. Of course, the investigator may or may not be even close to having really random samples and nobody knows how close he is. If we grant for the sake of argument that his sampling is random, it is obvious that in the long run he would reach the objective probability, the limit of the series, because this type of inquiry supplies just the right data for the successful functioning of the estimator; and this is what is meant by saying that this method is self-corrective. But although this method is self-corrective, it is self-corrective only under the assumption of ideal sampling plus the assumption that the long run will be reached. This is embarrassing because the long run is never reached. The procedure is not self-corrective under even ideal sampling conditions in the short run, for the limit as predicted at any given time may not be the long run limit. Though our estimates will in principle self-correct, we have no way of knowing when they are self-correcting, when they are merely static without centering around the limit, or when they are misleading relative to the limit.

The situation we are in is this: we are at all points following a procedure which we feel is reasonable or rational. Yet, when we look at the question of whether or not we are attaining truly random samples, we find that we cannot show that these samples are random. The best that we can show is that we are trying to attain random samples. And when we look at our sample average with the purpose in mind of comparing it with the limit, we see immediately that the limit is something which we are only trying to get, not a thing which we have now or ever will have before us, leaving us with no comparison to make.

We *could* compare earlier estimates with present estimates and restrict our generalizations to what has so far in fact happened; and we *could* state that we tentatively accept the present average as *the* population average, i.e., as the limit which in principle would be reached in the long run. But Peirce came to see that qualifications such as *experientially* and *provisionally* did not really demonstrate the reasonability of his frequency theory of objective probability.

It is difficult, at this point, to state just what position we are in. On the one hand we feel that our procedure is reasonable; on the other hand, we cannot show that this procedure is, in fact, producing the truth which is, in Peirce's view, the limit we would obtain in the long run. It may very well be the case, and Peirce does think that this is the way the problem stands, that we must at this point either abandon altogether the view that the sampling procedure is reasonable or else discover something else upon which to rest or ground the view that the procedure is reasonable. Peirce suggests that in the absence of any alternatives we must, in the end, ground this general theory of objective probability on nothing more than a feeling, i.e. on a sentiment. However, this sentiment is, Peirce suggests, a *logical sentiment.* Because of this sentiment, we are interested in an indefinite extension of scientific inquiry, we are interested in an unlimited community of scientists, and we feel that other human beings should feel the same way. We will return, later, to consider whether or not and in what sense such sentiment is reasonable.

So much for our account of Peirce's frequency theory. This account has been given with two purposes in mind: the first is to put enough of Peirce's frequency theory of probability before the reader to enable him to understand and appreciate Will's two criticisms; the second is to set the whole matter in a clear enough light to enable the reader to see, with only a modicum of additional comment, the weakness of Will's two criticisms.

• Will's two criticisms of Peirce's frequency theory of probability. • We will state, first of all, as succinctly as we can, Will's first criticism. According to Will, an adequate account of the meaning of probability must be an account of the meaning of probability expressions as these are employed in ordinary language. And, Will states, "... the primary function of probability expressions is the two-sided one of specifying the force appropriate to assertions actually made or considered and of indicating therewith the quality of the grounds available for these assertions." (op. cit., p. 65) Will observes that expressions like "A probably" or "A is probable" or "It is probable that A," convey something about the preferability of A by way of belief. What is conveyed concerns the force of the assertion A (actually made or entertained) and the quality of the grounds for asserting A. Specifically, such expressions indicate that the grounds for A are such that it is reasonable to believe A. For example, the assertion, "It will probably rain," conveys that it is reasonable to expect rain. In an ordinary case of someone's using a probability expression, to ask him if he thought his assertion reasonable would be to ask him to repeat what he had already said because the function of his probability expression in the first place was to claim reasonability. Now Peirce's account (and any other account) of probability which allows the question of the preferability or reasonability or rationality of what is already probable to arise must be a false account; for such an account obviously has not noted that the probable is the reasonable. To this, his first criticism, Will adds the suggestion that one of the reasons Peirce went astray is that he tried to define what is essentially epistemic in nature, *probability*, in terms of facts.

Will's second criticism concerns Peirce's suggestion that there is a logical sentiment, or, as we have interpreted Peirce, that a certain sentiment is, in itself reasonable. Will's criticism, in this case, can be put quite succinctly: Will says that this move by Peirce is bizarre. In the general procedure of sampling and estimating, we are not striving for a limit which we will reach, but only for a limit which we would reach if we continued the procedure indefinitely. We never do reach that which we only would reach, we do not know how close we are, etc.; and so on the basis of these and other shortcomings, Will points out, Peirce was forced to make the bizarre move of trying to defend the reasonability of the frequency theory by grounding it on a feeling which is supposed to be, in itself, rational, i.e., on a logical sentiment.

• Defense of Peirce's frequency theory of probability. • Both Peirce and Will agree that *probability* is frequently used to mean *reasonability*. Will's first criticism seems to be unfair to Peirce because Will accuses Peirce of raising the question of the reasonability of probability in the sense of reasonability when the question which Peirce does raise is that of the reasonability of objective probability.

What Will would have to prove, in order for his first criticism to hold up, is that objective probability insofar as it is different from untechnical probability, never occurs or is not real, or something like that, since, as things stand, objective probability is real and it is certainly meaningful to question the reasonability of the general sampling procedure which in effect defines objective probability. It does not seem likely that Will or anyone else could make that sort of probability which has reference to physical facts disappear. Even if we were to approach the problem in a purely linguistic manner, which seems *prima facie* to unnecessarily restrict the range of our study, we could not show that objective probability does not occur. I can, right now, *say* that this coin (which I just flipped and set on the table) will probably be heads, and that with a probability of 1/2; now it cannot be denied that I have objective probabilities in mind, though in addition to this I might well also have in mind probability in an untechnical sense.

A more substantial point at issue is whether or not Peirce is correct in defining reasonability in terms of certain kinds of facts. Will does not discuss Peirce's theory of reasonability at all, but he does assert that reasonability cannot be defined in terms of facts. It seems clear from Will's article that the kinds of facts Will has in mind are physical facts, and that the arguments which he has in mind (but does not give) which show that reasonability cannot be defined in terms of facts are arguments which hold up if and only if it is physical facts upon which the frequency theory of reasonability is based. But as we have seen, Peirce's frequency theory of reasonability is based upon both physical and cognitive facts with emphasis upon cognitive facts, albeit vague cognitive facts. Therefore Will's criticism is misdirected.

The most substantial point at issue, relating to Will's first criticism, concerns Peirce's frequency theory of reasonability. Both men have agreed that probability frequently means reasonability; the difference is that Will presents no theory about the nature of reasonability whereas Peirce presents a frequency theory; the substantial point at issue is whether or not reasonability does depend upon frequency, frequency in any one of the several respects mentioned above and in the qualified ways mentioned above. The minimal requirement, it must be remembered, in order for the frequency theory of reasonability to hold up, is that at least one analogy is involved in every instance of reasonability. Whether or not this is the case can, in principle, be decided by observation. On the one hand, when we reflect, when we look back upon the arguments we accepted and our reason for accepting them, we do remember analogous arguments. On the other hand, we occasionally seem to pronounce some arguments correct immediately, without reflection; and upon reflection we *may* not be able to remember an analogous argument. This could be explained by arguing that we had, on the basis of previous occurrences of the type of argument in question, formed a habit such that we immediately, by habit, believe the argument to be correct. This *could*, of course, be begging the question at issue; but the notion of a habit is certainly reasonable *qua* hypothesis since most of the arguments the correctness of which we claim to have immediately grasped are arguments of a type which occur all the time.

Perhaps a more convincing line of argument centers around the question of whether or not, when frequency is absent, reasonability is absent. Peirce was convinced that if frequency were absent, then reasonability could not be present. What this means, in terms of minimal requirements, is that if one had no previous instance whatsoever, if one had no opportunity to compare the argument in question with at least one analogue, then one had no reason for deciding for or against the argument or event in question. Peirce favored the materialistic view of probability over the Laplacean or conceptualistic view for just the reason that the conceptualistic view considered itself to have reasons for things where frequency, the experience of at least one previous case, was not at hand. In order to apply the probability calculus to specific material, the conceptualists had to assign each event an initial probability. In cases where the conceptualists could not make an initial assignment (based upon experience and estimation of the chances involved), in cases, i.e., where the probability of an event happening might possibly be anything from zero to one, they chose 1/2 as the initial assignment, a choice which is obviously arbitrary. It is clear that as the material which we survey in order to make our estimates may be physical or cognitive or both (in a polar relation), the same sort of argument would apply to the task of assigning initial probabilities to novel and unique arguments or argument types.

We come, now, to try to defend Peirce against Will's second criticism. In order to defend Peirce against this criticism, all we *have* to show is that it is *not bizarre* to hold that sentiment can be rational.

Now Peirce claims that it is rational or reasonable to follow a procedure and to take an interest in the completing of this procedure even though we can see from a strictly logical point of view that we will not reach our theoretic goal.

Peirce claims that, even in light of the strictly logical shortcomings of the frequency theory of objective probability, it is, nevertheless, reasonable to continue to follow the general sampling procedure. Scientists do consider the general sampling procedure to be reasonable and it is important to notice that there are procedures which are like the general sampling procedure in that they are at once considered reasonable and yet do not form a strictly deductive unity or, even, a defensible, nearly rounded out, inductive pattern, i.e., it cannot be shown that they are *approaching* deductive certainty. An example is Mill's method of residues. Generally, Mill's method would work, in the sense that they are deductions or approaching deductions, if we had a particular kind of universe. This universe would have to be such that in it existed a one-to-one correspondence between causes and effects such that each effect had one and only one cause, that these correspondences were everlasting, plus, perhaps, a few more qualifications. In this imaginary universe, the method of residues would be infallible. If ABC and only ABC caused abc and if A caused a and B caused b, then we would be certain that C was the cause of c. We do not have such a universe even in our best controlled experiments; even in the case of experiments in which we think we have a limited and identified number of factors involved and controlled, we are not sure of the total number of factors involved or of our identifications. And we cannot suppose a one-to-one cause-effect correspondence; unfortunately, we must always at least anticipate a plurality of causes. *Yet*, if the scientist has attempted to meet those conditions which would have to be met in order for him to isolate the cause of an effect (a particular functional relationship), with some further easily understood qualifications, then we consider his procedure to be reasonable. While we initially may have thought that the scientist is isolating the *cause*, we admit, upon reflection, that his results must be tentative; and when pushed further, we are forced to admit that we only *feel* that the procedure is reasonable. There are, of course, a multitude of grounds for our counting these scientific procedures rational or reasonable. These procedures give useful results, they give increasingly better estimates, they are consistent, usually, with one another, etc. But we still are brought back, in the end, over the same ground and must admit that in the end that which we must show to be reasonable is nothing more than a certain sentiment.

Our belief in this matter is that there is a sentiment in the case of the sampling procedure and that this sentiment is reasonable. Unfortunately, the only argument we can give for this belief is to point to experience. This sort of defense is similar to the defense of ordinary language philosophers: when asked to defend their assertions, ordinary language philosophers point out that such and such is the way we use language. We point out that such and such is the way we feel about the sampling procedure. The difference is that we cannot easily identify the sentiment; whereas ordinary language philosophers *are able* to discover identifiable ordinary uses of language.

Human beings (mostly scientists perhaps) who are acquainted with the sampling procedure and who *feel* that the procedure is reasonable think that other human beings ought to have the same sentiment; those who have the sentiment are of the opinion that their having this sentiment is not and should not be peculiar to them alone. In short, it is a fact that those acquainted with this sentiment think that this sentiment is universalizable. If what is universalizable is reasonable, then it follows that this sentiment is reasonable.

The argument given here is then, simply, that the sentiment which lies at the root of the frequency theory of objective probability is universalizable and therefore, if what is universalizable is rational, rational. This argument is intended to mirror the Kantian argument which leads to the antinomy of taste. In the *Critique of Pure Reason* Kant deals with the problem of how synthetic *a priori* cognitive judgements are possible; in the *Critique of Judgment* he deals with the problem of how synthetic *a priori* esthetic judgments are possible. Kant points out that some of our feelings are not universalizable and therefore not rational. Presumably those of our feelings which are universalizable (among other things) are therefore rational. Those arguments which we expect to hold in all cases and which have held in all cases, we pronounce valid. Similarly, Kant maintains, those sentiments which we expect to hold in all cases we pronounce valid. Kant calls this subjective validity, general validity, esthetic validity, and common sense (meaning common sensibility). While esthetic judgments are like strictly logical judgments in that they are universalizable, they are unlike strictly logical judgments in that we can give no further delineations. The way Kant puts it is as follows:

> The principle of taste, therefore, exhibits the following antinomy:
>
> 1. *Thesis.* The judgment of taste is not based upon concepts; for, if it were, it would be open to dispute (decision by means of proofs).
>
> 2. *Antithesis.* The judgment of taste is based on concepts; for otherwise, despite diversity of judgment, there could be no room even for contention in the matter (a claim to the necessary agreement of others with this judgment). (*Critique of Judgment*, section 56)

While the Kantian perspective takes an interest in various faculties and in *a priori* synthetic judgments, our perspective is more empirically oriented. This means that while we have mirrored Kant's argument about taste, we would not want to claim with him absolute universality or necessity in regard to the sentiment we are investigating. We accept the fall back to a more Humean position. This means that we look at the universalizing as an observation that scientists for the most part agree about the sentiment and expect others to agree. This leads to the interesting speculation that a sentiment lies at the root of the frequency theory of probability, that that sentiment is reasonable, and that the reasonability of that sentiment depends, in turn, and once again, upon frequency. Thus, if this speculation is correct, Peirce need not necessarily abandon the frequency theory to logical sentiment *simpliciter*, but to a sentiment that is rational because and in so far as it, in turn, is based upon frequency (which is circular, but not a viscous circle).

But there is another respect —a less theoretical and more practical respect— in which the logical sentiment in question is reasonable. The state in which we are accustomed to finding ourselves is in part stable and in part unstable. In regard to some things we have no doubt; in regard to others we do doubt and we do very reasonably inquire so as to replace this doubt by belief. At each point in our experience or in the history of science we inquire so as to replace relatively unstable configurations by relatively stable ones. According to Peirce's guess, the general course of this evolution is from a primordial state of relative chaos to a state in the long run of perfect crystallization of thought, i.e., in the long run we will settle down into an overall stable state, in *fixed* belief in a sense analogous to chemical fixation. The point in this evolution at which we nowadays find ourselves is analogous to a point midway between disorganized configurations of matter and the organized configurations of crystal: in short, the relations between our thought-signs is analogous to the nature of glass, a material which is partly disorderly and partly crystalline. In the long run it will no longer be practically reasonable to doubt our procedure because *all* real doubt will have ceased. One practical reasonable feature of our present procedure is that it is the only way we can move toward the time when all real doubt will cease. The alternative is to block the road of inquiry.

Glendon College, York University,
Toronto, Ontario.

Language and the Self

by Duane H. Whittier

Language and the self are two topics which have received considerable attention from recent philosophy, both analytic and phenomenological. What has been relatively neglected is the topic of how language and the self are related. In my opinion the most insightful things on this subject have been said by C.S. Peirce, and the purpose of this paper is to attempt an elucidation and elaboration of these Peircian insights.

Peirce asks the question: What is man? His reply is that man is a symbol. At first glance this answer seems eccentric in the extreme. However, I am convinced that it is literally true. When we ask what a man is, we want to know

> To what real kind does the thinking, feeling, and willing being belong? We know that externally considered man belongs to the animal kingdom...; but what we seek is his place when considered internally, disregarding his muscles, glands, and nerves and considering only his feelings, efforts, and conceptions.(1)

Peirce holds every state of consciousness to be an inference and life a sequence of inferences. Thus at any given moment man is a thought, and "as thought is a species of symbol, the general answer to the question what is man? is that he is a symbol".(2) To pin-point the meaning of this assertion he suggests we compare man with some other symbol. He uses the word *Six*. In what respects does a man differ from that word? He finds only two: (1) "the body of man is a wonderful mechanism, that of the word nothing but a line of chalk"; and (2) "the meaning of the word is very simple, the meaning of a man a very Sphinx's question".(3) But what other differences are there in addition to these two obvious and surface differences?

Peirce first examines the view that man has consciousness and a word has not. He notes that we must inquire about consciousness, for it is an ambiguous term. He distinguishes between what I shall call "organic consciousness" and "symbolic consciousness", "organic consciousness" being "A consciousness which is dimmed when animal life is at its ebb, in age or sleep, but which is not dimmed when the spiritual life is at its ebb; which is more lively the better animal a man is, but is not so the better man he is."(4) Let me elaborate this distinction. I think we must distinguish three levels of clearness in awareness (clarity of consciousness): (1) Consciousness as wakefulness or "cerebral alertness" —this being a matter of intensity (threshold), dependent upon organic conditions, and something which ebbs with fatigue and sleep; (2) awareness as involving a mental set or disposition to attend to (focus upon) something —this involves the matter of concentration *versus* distraction; and (3) consciousness as heightened awareness —something that is a function of the wealth of relations perceived, the ingathering of various elements for simultaneous apprehension. Meerloo puts my meaning well when he states: "The very symbolic function is a time-gaining device. It is a throwing together (symballein) of several visions in a simple written or painted sign. Grasping symbols is grasping a condensation of human experience, of human time."(5)

Let me further explain (3). In congnition, one the better understands and appreciates something the fuller the background context of relations. The more relations we can behold X in, the greater our awareness or grasp of it (i.e. the more sharply *delineated* X is, and the better able we are to *articulate* it). For example, your awareness of some truth is heightened if you can grasp it against a background of previous error on your part concerning it. Your awareness of the meaning of some scientific law or principle is greater if you can give ten different illustrations of its application rather than only two. As we say, you have a greater "feel" for what is involved. (Keeping in mind Meerloo's view that a word is a condensation of human experience, one can see why I consider these two examples to be illustrations revelatory of the nature of "symbolic awareness".) Clearly awareness in the sense of (3), "symbolic consciousness", presupposes awareness in the sense of (1), "organic wakefulness", and (2), "proper mental set", but not *vice versa.* There can be maximum wakefulness and maximum concentration or focus and yet a poverty of awareness due to lack of *conceptual* grasp. In no way is awareness in sense (3) a function of physiological or psychological factors. It is a function of semantic factors, of words and their relations. Holding the intensity (degree of wakefulness) and focus (mental set) of consciousness constant, the question as to the quality (clarity) of a state of consciousness can still be raised, and it can be answered only in terms of considering the deportment of the language of the mind concerned. Consciousness is always *consciousness-of* something, and the quality (clarity) of the state of consciousness (still holding wakefulness and focus constant) is a function of the design or intelligibility inherent in the relations holding between words. In this case the *clarity of consciousness* and the *articulateness of the content in consciousness* are one and the same thing. This is why that unjustly neglected American philosopher, Elijah Jordan, can in all literalness say about a poem, that it "is not merely felt by me or you but felt absolutely and by itself; it feels itself, is feeling; *consciousness of itself develops within* it, out of its own synergy...."(6) It is the deportment of words in the poem that gives us that lucidity the poem confers. On Peirce's view both what I call "organic consciousness" and "psychological set" belong to man as an animal, but what I term "symbolic consciousness" belongs to man insofar as he is a self. Thus selfhood without language is impossible. The self is its language. The peculiarly human form of consciousness is through and through linquistic in nature.

Peirce notes that "consciousness" is also used to denote feeling. Do words have feeling? Clearly not. But do not words have something corresponding to feeling? Keep in mind that "feeling is cognitive" (by which Peirce means that all feeling is feeling-*of* something, i.e. no feeling *merely feels itself*). Man's feelings are perceptions, he is influenced by objects. He sees, hears, touches, etc. A word does not do these things. Yet there is here

> a correspondence between the word and man. Perception is the possibility of acquiring information, of meaning more; now a word may

learn. How much more the word *electricity* means now than it did in the days of Franklin; how much more the term planet means now than it did in the time of Hipparchus. These words have acquired information; just as a man's thought does by further perception. But is there not a difference, since a man makes the word and the word means nothing which some man has not made it mean and that only to that man? This is true, but since man can think only by means of words or other external symbols, words might turn around and say, You mean nothing which we have not taught you and then only so far as you address some word as the interpretant of your thought. In fact, therefore, men and words reciprocally educate each other; each increase of a man's information is at the same time the increase of a word's information and *vice versa.* So that there is no difference even here.(7)

I think Peirce in the foregoing forces us to come a long way toward agreeing with him that "remote and dissimilar as the word and the man appear, it is exceedingly difficult to state any essential difference between them except a physiological one."(8)

Elsewhere Peirce states that "There is no element whatever of man's consciousness which has not something corresponding to it in the word; and the reason is obvious. It is that the word or sign which man uses *is* the man himself. "...my language is the sum total of myself; for the man is the thought."(9) A.M. Adams is the only author I have discovered (in addition to Elijah Jordan) who seems to share Peirce's view here. Adams states:

Perhaps the best analogue of the relation between the mental and its physical foundation is found in language. It has both a physical and a semantic dimension. The physical ink marks on paper or noises in the air do not constitute language, but there could be no language without them or something physical with the same function. A language is like a mind in other respects also. It is not only an instrument to be used in mental performances, but it has the structure of mind. We think of things; language contains denoting words and phrases. We think that such and such is the case; language contains declarative sentences. We question; language contains questions. We feel constrained to do something; language contains 'ought'-sentences. We want things; language contains imperative sentences, etc. *We might say that a mind is a language which functions on its own.* It denotes things, it describes them, it asserts facts, it questions, it prescribes, etc. It means in all the ways in which language means. The neurological events and muscular activities involved are to the mental what ink marks or noises are to written or spoken language.(10)

I agree with Adams and Peirce that the mind or self cannot be reduced to the body. Mind is not identical with the brain. My simplest reason for so agreeing is one that appeals to a fact about the nature of language itself, namely, the fact that it is tensed. In language (and thus in thought) we have past-tense and future-tense (to take only these two), and it certainly is difficult to understand how these tenses could be rendered in purely physiological terms. That the difference between two sentences, a past-tense sentence and a future-tense sentence, can be represented by the difference between two physical states seems quite unintelligible. How could tense be a physical property of a physical state or process? What could the difference possibly look like in *solely physical terms*? That there are neurological differences reflecting the physical differences in the sound or visual appearance of words is readily intelligible, but that meaning-forms are physical is not. As I shall show below, reducing thought to language is very plausible, but reducing thought or language to physical process is impossible. There is no physical relation with which a semantic relation could be *identical.* The relations that can be exhibited by physical things are exclusively spatial in nature, and tense differences are not differences in spatial relations. At best one would have to say that there is a physical *correlate* of tense, or that physical processes can have non-physical *properties.* But if one does this, one is asserting Adams' position, rather than mind-body identity. Adams does not assert that mind or language can exist independently of the physical. The physical is a *necessary condition* for the *existence* of mind or language, but it is not in any way *explanatory* of the *nature* of mind or language.

Relative to this matter of language and the mind-body relation, Whitrow has written some very interesting things. Using memory as a case in point he asks, "Why do we suffer from *total* amnesia concerning the events of early childhood? " Freud, the first to consider the problem, advanced the idea of a subconscious "censor" which suppresses from consciousness the recollection of "infantile sexuality". Whitrow finds E.G. Schachtel's answer more convincing: "childhood amnesia results from a time-lag in the development of the conceptual and conventional memory schemata which are ... necessary for conscious recall." He continues:

> Memory is thus found ... to be essentially related to what Suzanne Langer calls *man's symbolic transformation* of experience. As she herself has suggested, the origin of the concept of 'self' —which is usually thought to mark the beginning of actual, that is, conscious, memory— may well depend on the process of epitomizing our feeling symbolically. "To project our feelings into outer objects is the first way of symbolizing and thus of *conceiving* those feelings." However, a symbol is usually very different in nature and appearance from the thing it stands for. Here, I believe, we have a clue to the curious fact that anything which has been, or is likely to be, suggested as the cerebral trace... of a memory ... seems to be utterly different from that which originally gave rise to it: whatever physiological mechanism may be involved, *the mnemonic trace functions for us as a symbol.*(11)

The reprocessing of raw experience through "symbolic transformation" is precisely that process out of which the self emerges. Non-verbalized experience is "unconscious", i.e. it is unrecognized, and not in a form to be grasped and worked with. Language makes the absent present to us, but also "distances" that which is present. Non-verbalized experience is not "distanced" by language and is thus too close to us to be reacted to in a voluntary (responsible) way. Animals have no language, and because of this no "free will".

Whitrow defends his suggestion that the mnemonic trace functions as a symbol —"this interpretation throws light on the otherwise puzzling fact, to which W. Riese, himself a physiologist, has drawn attention..., that recognition of the *symbolic* nature of thought and its linguistic expression has been found 'to promote greater insight into the dynamics of speech defects resulting from brain lesions than their description in merely physiological terms, those of purely motor or sensory type' ".(12) Indeed, all these considerations strongly support the view that it is unlikely that our capacity for recall-memory can be explained in mechanical or chemical terms. And Whitrow's discussion of the symbolic nature of memory processes lends support to the Peircian view that the self is identical with its language, and where we have a human being not yet in possession of language, we have a creature minus a self, a creature who has no more than the "organic consciousness" animals have.

Peirce advanced his view of the identity of the humanly conscious self and language in the context of discussing immortality. It is not my purpose here to discuss immortality, but to try to show the fruitfulness of his view by examining a problem which I think it solves —that of human communication. In preparation for doing this I must present a fuller articulation of his view with an eye toward making it more plausible. I think the best way to start doing this is to contrast it with the view to which it is radically opposed. "Mentalism" is the doctrine that words symbolize mental experiences which in turn symbolize external things (words thus indirectly symbolizing external things). For example, the word "tree" *stands for* my mental concept of a tree which in turn *stands for* the tree itself. If "mentalism" is correct, then the act of communication, of using words, commences with the occurrence of a non-linguistic mental act of meaning or intending. This act finds outward expression in the selection and utterance of a sequence of sounds (the "linguistic sign") which is understood because of the conventions of the speech community which determine just what things the speaker's words are to be interpreted as referring to. Communication is thus viewed as the process of encoding and decoding a pre-existent, non-linguistic, "mental message". Fodor calls this view the "replacement theory" of meaning, and I think this expression preferable to "mentalism". With respect to it Fodor states:

> Notice that it is a consequence of this model that the nature of the message —that is, of the mental message or "idea"— is or may be completely undetermined by the characteristic of the telegrapher's code. The act of meaning a thing is distinct from the act of uttering a sentence in just the way that filling out a telegraph blank is distinct from encoding it.(13)

Peirce in effect rejects the "replacement theory", for on his view everything about the human mind that is not shared with animal minds is not *represented* by language, but rather is *constituted* by (is identical with) language. Certainly the introspective evidence on this matter seems unambiguously on the side of those who reject the "replacement theory". Witness the words of Wittgenstein:

> What happens when we make an effort —say in writing a letter— to find the right expression for our thoughts? —This phrase compares the process to one of translating or describing: the thoughts are already there (perhaps were there in advance) and we merely look for their the question: "What did the thought consist in, as it existed before its expression? "(14)

> When I think in language, there aren't 'meanings' going through my mind in addition to the verbal expressions: the language is itself the vehicle of thought.(15)

Now if we hold that language *constitutes* the uniquely human forms of mental experience rather than being merely an encoded representation of it, we are committed to what seems, at first sight, a most implausible view. How can the descriptive use of language (language that describes or designates my mental experiences) be accounted for on a theory that all language is constitutive of humanly conscious experience? The answer is to show that the descriptive use of language is dependent upon and presupposes the prior constitutive functioning of it. Think on this fact: to a doctor a symptom is a sign of the illness. To the patient it *is* the illness, or part of the illness. This fact suggests the possible relation between the constitutive and descriptive (referential) aspects of language. Just as fever and rash can be signs or symptoms of measles to the doctor, simply because they are, for the patient, constituent parts of the disease, so also "I hope she will come for Christmas" can be taken by the external observer as a *report of* or *reference to*, my mental state only because for me "I hope she will come for Christmas" is not a reference to or sign of my mental state, but rather *constitutes* it. (Can anyone seriously hold that *hoping* as a mental activity could occur in the absence of language?) A word, to be a sign of a mental experience, must be a *constituent part* of that experience. Words must give form to, and be constituents of, a particular content before they can be used to communicate such content. As E. Jordan remarks: "Where a word as symbol stands for an object no part of which it is, there is the breach that never can be filled and which cuts the world in two forever."(16)

Linguistic meaning is an objective or public phenomenon, and not something locked up in the inner recesses of private minds. That is, meaning belongs to words as public objects, and not to minds with words merely serving as "signs" for the meanings contained therein. This is to say that the rejection of the "replacement theory" commits one to the view that words are primarily related to one another and not to minds, and this means that the communication process involves word-to-word rather than word-to-mind relations (transactions).

We must examine precisely what this means. O.H. Mowrer notes how "widely accepted is the notion that the basic function of language is to transfer or bear meanings from person to person, from mind to mind."(17) He notes that the general thesis takes the form "Language is the expression of thought by means of words" and "The chief business of language is to communicate meanings of various kinds." Then he goes on:

> Let us explore now, instead, the proposition that in communication we are not transferring meanings from person to person as much as we are transferring meanings from *sign to sign* within a given person, within a single mind. Said a little differently, the suggestion is that in order for us to communicate effectively with another person, he must already *have* the meanings with which we shall be dealing and that in the communicative act we are, for the most part, merely changing the signs to which particular meanings are attached, merely shifting or transferring meanings from one sign to another. One person, by the use of

> appropriate words or other signs, can *arouse*, or "call up," particular meanings in the mind of another person; but he does not "transfer" or implant them there. The communicative act, in its most salient aspect, lies rather in the combination, juxtaposition, or association of the meanings thus aroused in *novel, "informative" ways.*(18)

The thesis here is that communication is primarily an affair of words or meanings "interacting," and derivatively from this a mind-to-mind relation of meeting and sharing. Mowrer's thesis accords well with experience. For example, take the situation where we become acquainted with the thought of a previously unknown philosopher. We become familiar with unknown thoughts and new ideas only through his words. Think, for example, of the Kantian world of thought, or that found in Aquinas, and contrast it with the world of thought to be found in a Hegel, or that presented by a Heidegger. In addition to obvious neologisms, we encounter in the works of these different men all kinds of familiar words which do not have the familiar meaning (a fact we may not notice initially). As we read on, we gradually come to notice the new way of speaking and the new meaning of words. We begin to grasp the cast, and appreciate the flavor, of the author's peculiar style of thinking. We become acquainted with his new way of thinking only through the medium of words. We must, I think, conclude that thought is really present in words. That is, *to learn someone's language is identical with penetrating into his thought.*

When Mowrer speaks of "novel, 'informative' ways" he is pointing out that communication is neither a responding to new signs with old responses, nor a matter of simply having familiar symbols aroused and left at that. Communication involves, in its most significant aspect, making and attending to something new. And I have tried, in the above paragraph, to indicate how it is that a new way with words is identical with a new way of thought. The structure of a language makes it possible for us to understand statements we have never heard before and to say things we have never said before. New meanings arise out of the interrelating of old ones, but we must fully recognize that such combinations are not just eclectic affairs, rather the synergy of words is such that the qualitatively new can emerge from such combinations.

If the foregoing lines of thought are correct, then it can be seen that communication is not a physical-causal process at all. On the physical-causal level I utter sounds, which are transferred through the air, and you hear them. But sounds do not literally carry meanings in a physical sense. *Communication is an affair, not between signs and minds, but between meanings.* One mind is able to communicate a new idea to another solely by virtue of the fact that words (as fields of connotations) can exhibit new meanings when *they* meet and flow together.

We must distinguish between two kinds of "communication": (1) the kind of communication we have when an animal gives out a mating call, or cry of warning, etc. This is *communication as signaling.* At the human level this kind of communication takes the form of language viewed as a tool for obtaining social concert. This is *communication as interaction*, we are said to be "in communication" if we are in interaction. (2) What I term "genuine communication" occurs when two minds meet and regard the same content. Here we have language as *communion* rather than as merely an instrumentality. The type of communication under discussion in this paper is communication viewed as communion of minds (i.e. shared contents) rather than viewed as a tool in the interaction between selves engaged in some practical operation or physical task.

How do the meanings which never leave my mind (in the physical-causal sense) come to "correspond" to yours? The answer is simply that meanings never were private. It was never the case that we had them privately and then acquired some public sign or symbol by means of which to communicate them to others. On the contrary, meanings come from without. You and I acquire (learn) them from the public realm in the same way. Private minds do not possess meanings prior to and independently of learning language, *rather the process of learning a language is the process of building or acquiring a mind.* Minds do not have meanings apart from having a language, and then seek word-vehicles for their outward (public) manifestation (this is the "replacement theory" of meaning), rather the word in its constitutive role (it constitutes thought itself) gives the meaning to the mind.

We, as individuals, do not *create* word-meanings. In the initial learning-stages of our lives we *acquire* them, and then later on, when hearing or using new meanings, we *discover* them (i.e. we discover how fields of connotations interrelate to produce new determinate meanings). School-teachers and dictionaries instruct us, and if we become poets, the very language itself instructs us. In the dim history of our race, word-meanings probably emerged initially out of social practice (ceremony and rite), then once a certain "critical mass" was achieved, language was able to breed new language from out of its own resources. There is no problem of how my mind with its private language (or non-linguistic thought) establishes contact with your mind with your private language (or non-encoded thought) so as to work out a common language by means of which we can communicate. It is not the case that our minds are self-formed and develop independently of the acquisition of a public language, thus giving us the insoluble problem of trying to work out a common meeting ground (which is *ex hypothesi* impossible, for one cannot meet to make arrangements for an initial meeting). The truth is that we all imbibe from, and are monitored by, the same language community as we grow up, and this means that our very minds appear as products of the language and are not something prior to it.

It cannot be denied that there is a responsibility in putting words together and "communicating," and it belongs to *each individual* using the language. I do not wish the foregoing remarks to be construed as denying or minimizing this responsibility. Nor do I wish to deny that communication sometimes fails, due to individual irresponsibility in the use of language, or due to ignorance. But the focal point of my interest in this paper is in what must be the case in principle in order for any communication to be possible at all.

I want to explore further (below) the thesis that I have been trying, by degrees, to justify above; namely, that communication is a relation between minds by virtue of being first of all a relation among words. By examining the topics of "behaviorism" and "solipsism" in the context of Wittgenstein's thought, I want to show that on the "replacement theory" of meaning genuine communication is impossible, and the only possible alternative to the replacement theory is Peirce's view that mind and its language are identical.

The traditional view holds that the language of sensations has a double aspect. One is public and related to physical behavior, the other is private and related to the private sensations of the individual. This is the sort of view that Wittgenstein, in his *Philosophical Investigations*, rejects. He denies the validity of the idea of a private or "hidden" experience, and argues for this by showing that it is impossible that language could come to refer to such private experiences. His basic claim is that an "inner process" stands in need of outward criteria. He states:

> What would it be like if human beings showed no outward signs of pain (did not groan, grimace, etc.)? Then it would be impossible to teach a child the use of the word 'tooth-ache'. —Well, let's assume the child is a genius and itself invents a name for the sensation! — But then, of course, he couldn't make himself understood when he used the word.(19)

It is because of what we might call "sensation-behavior" that a common sensation-language comes to be built up. If sensations were private experiences without "outward signs." it would be impossible to give them public names at all. Wittgenstein seems to imply here an *essential* connection between sensation and behavior.

> How do words *refer* to sensations? —There doesn't seem to be any problem here; don't we talk about sensations every day, and give them names? But how is the connexion between the name and the thing named set up? This question is the same as: how does a human being learn the meaning of the names of sensations? — of the word "pain" for example. Here is one possibility: words are connected with the primitive, the natural, expressions of the sensation and used in their place. A child has hurt himself and he cries; and then adults talk to him and teach him exclamations and, later, sentences. They teach the child new pain-behavior.(20)

Assuming Wittgenstein does hold that there is an *essential* connection between sensation and behavior, and holding as he does that sensation-language is an acquired form of sensation-behavior which substitutes for natural pain-behavior, it seems reasonable to conclude that he agrees with Jordan's principle that a word must be a constituent part of the mental experience it signifies, else there is a gulf that cannot be bridged.

Wittgenstein has been accused of being a "behaviorist." Whiteley has given a clear statement of the essence of behaviorism:

> Many philosophers of the past have treated mental concepts as though only their inner components were essential, the behavioral factors being merely accessories, attached to the concepts because a certain sort of behaviour normally accompanies the sort of experience which primarily constitutes the state of mind. This is an error, which not only misrepresents linguistic usage, but misrepresents the character of mental life as we perceive and understand it. [There is a] complementary error, that of supposing that mental concepts can be construed as essentially behavioural, *the private components being either ignored or regarded as dispensable accessories*; so that a man's being drowsy consists in his

behaving in a certain way, and the way he feels is irrelevant in determining whether he is drowsy or not. This is what I shall mean by Behaviourism.(21)

On my interpretation, Wittgenstein is not a behaviorist in this sense. For him, feeling and behavior have an essential connection. Yet, because of his stress on language as a doing-in-the-world, Wittgenstein's account does slight the qualitative or content aspect of language, and thus appears to some as a species of behaviorism.

Ernest Gellner is one who interprets Wittgenstein to be a behaviorist. He states:

> ...Linguistic Philosophy of mind does not deny the existence of warm mental-entities.... It merely argues that they are irrelevant, in as far as their presence is not essential to the use —and hence, the meaning— of those expressions which were once held to designate them. For instance, an angry man may or may not be experiencing an inner feeling of anger: What is essential to his anger is only that he be disposed to *act* in an angry way.
>
> It must be noted how this philosophy of mind depends on the theory of language: if the meaning of terms is their *use*, as tools in the public world, then, naturally, the use of the tool can only be specified in terms of what is necessary to its employment. (If inner states may sometimes be applicable, then those states are not necessary for its use and cannot be a part of its meaning.)(22)

Gellner holds that Wittgenstein, curiously enough, has landed in solipsism by virtue of his very efforts to refute it — "he reached solipsism from a premise diametrically opposed to the one which normally leads to that position: he reached it from the view that language is necessarily public. Being a system of public acts, private experience was irrelevant to it."(23) Gellner is saying that Wittgenstein, by using purely behavioral criteria for licensing the use of "mental" terms, has dropped the content element out of sensation-language. If this were true, then indeed Wittgenstein would have shown solipsism out the front-door only to have let it in the back-door. That is, he would have so divorced language from the mental life of its users that we would have the spectacle of a public language and private selves with whom this public language fails to make contact. But Wittgenstein, by asserting that feeling has a behavioral dimension, and essentially so, is able to appeal to behavioral criteria and in so doing automatically to take into consideration feeling-content.

Gellner cites no reference, but the "beetle in the box" analogy may be what he has in mind as indicating Wittgenstein's commitment to behaviorism.

> Now someone tells me that *he* knows what pain is only from his own case! —Suppose everyone had a box with something in it: we call it a "beetle". No one can look into anyone else's box, and everyone says he knows what a beetle is only by looking at *his* beetle. —Here it would be quite possible for everyone to have something different in his box. One might even imagine such a thing constantly changing. —But suppose the word "beetle" had a use in these people's language? — If so it would not be used as the name of a thing. The thing in the box has no place in the language-game at all; not even as a *something*: for the box might even be empty. —No, one can 'divide through' by the thing in the box; it cancels

out, whatever it is.

> That is to say: *if we construe the grammar of the expression of sensation on the model of 'object and designation' the object drops out of consideration as irrelevant.* (24)

Wittgenstein denies that sensations are private, but he is here arguing that *if* there should be an utterly private sensation it could play no role in the language of sensation. *If* we construe the grammar of the expression of sensation on the traditional model of "object and designation" (where the sign is not continuous with what it signifies), then the object (pain) drops out as irrelevant, because it would be utterly private and thus could not be publicly named. But Wittgenstein has shown a way in which language and feeling can be related which does not follow the model of "object and designation" as set forth on the "replacement" theory of meaning. For Wittgenstein, natural sensations (e.g. pain) have a natural behavioral dimension (e.g. crying). In learning language we are taught new (verbal) pain-behavior which substitutes in part or entirely for natural pain-behavior. The process is roughly analogous to the way fallen wood can be gradually replaced by calcium until we have "petrified wood." (This example can serve as a paradigm of how "symbolic consciousness" is related to "organic consciousness.") Verbal behavior becomes related to the overall pain-experience (i.e. pain-feeling *plus* pain-behavior) as a *constituent* of it. And the later ability to use language to refer to pains and report them is *derived from* this prior and more basic relation. Without this more basic relation, pain-words could not publicly refer at all. Gellner misleads himself into thinking that Wittgenstein is denying the *relevance* of felt-content, when all he is doing is denying its *privacy*. All Wittgenstein is presupposing is that no public language can be built up out of private ones. (If my understanding of Wittgenstein is incorrect and Gellner's correct, then he has indeed made a devastating criticism of Wittgenstein on the matter of solipsism.)

Wittgenstein's account of language stops at a very simple level. A language-trained human organism experiences all sorts of subtly nuanced qualitative contents for which pain cannot serve as a paradigm. Some of these felt-qualities are not at all natural (i.e. they are not rooted in the pre-cultural physical nature of the organism, as pain is). For example, take poetic metaphors as felt-qualities, or the humorous quality of jokes. Metaphors and jokes are linguistic objects, they are not *signs* of anything nor are they symbols that *designate* anything extra-linguistic. In cases such as these, how does language stand to content? Pain-language does not communicate the pain itself: it communicates *about* the pain. (After all, to *describe* a pain is not to *suffer* a pain.) The words of a metaphor or joke are not related to their qualities the way pain-language is related to pain-feeling. Understanding pain-language may not constitute a pain, but understanding a live metaphor or fresh joke does indeed constitute a felt-quality.

Language not only constitutes the very stuff of certain *objects* (e.g. the aforementioned metaphors and jokes), it also makes possible certain *processes* which, in the absence of language, could not occur at all (e.g. hoping and death-anxiety). Man is not primarily a "symbol *using* animal," rather man *qua* man is *constituted* of symbols. As noted above in the words of E.M.

Adams —a man is a semantic field with a physical dimension. A particular self is an instance of an "automatic language."

There is indeed a weakness in Wittgenstein's account of language which Gellner may have sensed. Wittgenstein does slight the felt-content or aesthetic side of language in favor of its doing or functional aspect. In this respect his account is seriously incomplete and one-sided. But the important thing to recognize is that the denial of the "replacement" theory of meaning entails no denial of the aesthetic dimension of thought. It entails only the denial of the privacy of content (precisely what we want, if communication is to be rendered possible). It entails no reduction of thought to bodily behavior, only to just verbal activity. But verbal activity has its qualitative dimension. Mind is its language, but language is a *content* as well as a *doing*.

Peirce was not directly concerned with the problem of solipsism, but he did concern himself with the nature of man's "immortal soul". He states:

> Enough has now been said ... to show a true analogy between a man and a word. ... A man denotes whatever is the object of his attention at the moment; he connotes whatever he knows or feels of this object, and is the incarnation of this form or intelligible species; his interpretant is the future memory of this cognition, his future self, or another person he addresses, or a sentence he writes, or a child he gets. In what does the identity of man consist and where is the seat of the soul? It seems to me that these questions usually receive a very narrow answer. Why we used to read that the soul resides in a little organ of the brain no bigger than a pin's head. Most anthropologists now more rationally say that the soul is either spread over the whole body or is all in all and all in every part. *But are we shut up in a box of flesh and blood? When I communicate my thought and my sentiments to a friend with whom I am in full sympathy, so that my feelings pass into him and I am conscious of what he feels, do I not live in his brain as well as in my own—most literally? True, my animal life is not there but my soul, my feeling, thought, attention are.* If this be not so, a man is not a word, it is true, but is something much poorer. There is a miserable material and barbarian notion according to which a man cannot be in two places at once; as though he were a thing! A word may be in several places at once, ...because its essence is spiritual; and I believe that a man is no whit inferior to the word in this respect.(25)

Clearly, if mind is its language, then by virtue of this very fact, communication between minds is possible. We can assert that a man is *more* than his bodily behavior but even so his nature remains public.

To avoid any misunderstanding about the thesis that a mind and its language are identical, it is necessary that I straightforwardly answer this question: What is a word? To identify the human mind with words is absurd if words are only physical objects or events. To identify mind with words is redundant if words are in turn identified with something "in the mind" as distinct and separate from words as physical marks and sounds (for to do this would be simply to reaffirm the "replacement" theory). A word is not a physical mark or sound. A word is not a mental act as it might be apart from some physical mark or sound. Rather, a word is a physical mark or sound plus all the experiences and usages which are in-gathered around it and cumulatively embodied and crystallized about it. A word is like a pearl —it is a concretion of experience around the word-sound or mark taken as the grain of sand. Thus a word is not to be identified with some mental activity supposedly possible apart from a sound or mark. If a word is a concretion of experience, is it public or private? It is public. Experience as in-formed and contoured by the *rule-governed* and *socially monitored* use of sounds and marks is what constitutes a word.

It follows from this view that if one wishes to study the life of the mind one does not turn to the science of psychology. Not to behavioristic psychology, which studies behavior apart from felt-content. Not clinical psychology, which studies reactions rather than creations (i.e. a healthy creature is engaged in world-building projects, only the diseased turn attention in upon themselves and become focussed upon licking their wounds). Not neurophysiology, which, if it mistakenly thinks itself studying mind, is guilty of reducing mind to body, something metaphysically impossible. To study the life of the mind one studies the art, music, literature, law, politics, and religion of a people. The life of the mind is found embodied in these institutions and their respective languages. The life of the mind is not within, but without. It is not subjective and private, but objective and public.

SUMMARY. On the traditional or "replacement" theory of the relation of language to the self we are left with private centers of experience that can never make verbal contact with a public world. On the behavioristic interpretation of language we have a public language to which the private realm of experience is irrelevant. On both views the possibility of genuine communication between minds is either denied or left an utter mystery. On Peirce's theory of language communication becomes explicable. The self *is* its language. And, supplementing this with Wittgenstein's "private language argument", language is from its very inception through and through public. Thus to learn a language is to acquire a mind. There is no gap between minds to be bridged. Individuals are like branches from a common trunk, and like branches they are distinct but not ultimately separate.

When Jordan wrote: "It is barely possible that a structure of live words lying at the basis of the personality constitutes the 'mind',"(26) he was too modest. It is not "barely possible", it is clear fact, as Peirce saw.

NOTES

1. Peirce, C.S.: "Consciousness and Language" in *Collected Papers,* Vol. 7, (Arthur W. Burks, ed.), Harvard Univ. Press, 1958, para. 582.
2. Ibid., para. 583.
3. Ibid., para. 584.
4. Ibid., para. 585.
5. Meerloo, Joost A.M.: *The Two Faces of Man,* International Universities Press, N.Y., 1954, p. 102.
6. Jordan, E.: *Essays in Criticism,* University of Chicago Press, 1952, p. 37, italics added.
7. Peirce, op. cit., para. 587.
8. Ibid., para. 588.
9. Peirce, C.S.: "Man, A Sign" in *Collected Papers,* Vol. 5, (Hartshorne and Weiss, editors), Harvard University Press, 1931-5, para. 314.

10. Adams, E.M.: "Mental Causality", *Mind*, v.75, 1966, pp. 559-560.
11. Whitrow, G.J.: *The Natural Philosophy of Time*, Thomas Nelson & Sons, London, 1961, pp. 111-112.
12. Ibid., p. 112, footnote.
13. Fodor, Jerry A.: "What Do You Mean? ", *Journal of Philosophy*, v.57, 1960, pp. 499-500.
14. Wittgenstein, Ludwig: *Philosophical Investigations*, Basil Blackwell, Oxford, 1958, para. 335.
15. Ibid., para. 329.
16. Jordan, op. cit., p. 375.
17. Mowrer, O.H.: *Learning Theory and the Symbolic Processes*, John Wiley, 1960, p. 138.
18. Ibid., p. 139.
19. Wittgenstein, op. cit., para. 257.
20. Ibid., para. 244.
21. Whitely, C.H.: "Behaviourism", *Mind*, v.70, 1961, p. 164, italics added.
22. Gellner, Ernest: *Words and Things*, Beacon Press, Boston, 1960, pp. 24-25.
23. Ibid., p. 142.
24. Wittgenstein, op. cit., para. 293, italics added.
25. Peirce, op. cit., Vol. 7, para. 591, italics added.
26. Jordan, op. cit., p. 28.

University of New Hampshire,
Durham, New Hampshire.

The Critic of Institutions*

by Max H. Fisch

(*Presidential Address delivered before the fifty-fourth annual meeting of the Western Division of the American Philosophical Association at the Indiana University, Bloomington, Indiana, May 3, 4, and 5, 1956.)

The Greek name of philosopher, so proudly worn and so humbly —what besides the name have they had in common who in our western lands have worn it? We who sit here tonight as members of a philosophical accosiation—what have we in common besides the association? To press the question would be to call up more ghosts than could be laid again within the limits of your patience:—ghosts of particular and universal, existence and essence, member and class, substance and attribute, subject and predicate, real and nominal and persuasive defintion. One or two only of these ghosts I shall try to raise later, and let them haunt you if they still can, but I want first to take our bearings.

We are all familiar in a general way with the series of episodes in academic history by which the faculty of philosophy became a very loose collection of departments of arts and sciences, philosophy shrank to a single department within the collection, the arts and sciences multiplied, and numerous other professional faculties were added to those of law, medicine, and theology, so that philosophy is now no longer one of four, but one of a hundred, and no longer a prerequisite to the other three, but a competitor of the other ninety-nine.

When Locke wrote his *Essay*, he could look upon Boyle and Sydenham, Newton and Huyghens as the philosophers of the day, and profess himself to be only an under-laborer clearing the ground a little for such master-builders as these. Now it is Locke's *Essay* that is called philosophy, and nearly all that he called philosophy is science.

Since philosophy has shrunk from a faculty to a department, and diminished from a master-builder to an under-laborer, it might be expected, besides learning humility, to define for itself more clearly its now more limited task. Yet surely the diversity of what is now called, and of what calls itself, philosophy, is not less than it was in Locke's day, when he defined it as "nothing but the true knowledge of things." The questions we ask ourselves, or which, when addressed to us by others, we think properly addressed; the methods by which we seek our answers, or by which we justify them when found; and the tests to which we think it fair our answers should be put—have we philosophers as much in common in these respects as had those whom Locke called philosophers? Is there more than a family resemblance among us, bridging our extreme differences by likenesses of next to next, so that the extremes have in common only the series of which they are the opposite ends? Could we construct a definition of philosophy which, if it had authority, would not unchurch or unfrock many of the ablest among us?

Wittgenstein thought he had found a new subject or a new method which was different from what Plato and Berkeley had done, but which might be thought to take the place of what they had done, and might therefore be called philosophy. It is true that some of us now practice his method who might otherwise have practiced one or more of those already familiar and established. In any sense but that, however, his new method has not taken the place of any of the old ones, but has simply been added to them. The many have not become one, but have only been increased by one. So it has always been. Except for what has changed its name from philosophy to science, everything that has ever been called philosophy is still called philosophy, and, in spite of all changes of fashion, still survives in our midst.

May it all long survive! I have no wish to diminish the diversity or to make it seem less than it is. I am quite content that you should continue as diverse as you are, and yet all continue to bear the name I cherish for myself. If, therefore, I now proceed to recommend to you the oldest of all ways of thinking about philosophy, and to propose a new name for that old way of thinking, I ask you only to consider whether it includes what you are already doing and sets it in congenial relations with what other philosophers are doing. I shall be surprised and disappointed if it does not, but I ask no one to change his philosophic ways. If, however, there are younger members among us who have not yet found themselves philosophically, I suggest an arduous choice that will increase the diversity. And I have also a suggestion for departments of philosophy in our larger universities, which should have the same effect.

In spite of the vogue of analysis, perhaps the most familiar recent definition of philosophy is still that by Whitehead in *Science and the Modern World:*

> I hold that philosophy is the critic of abstractions. Its function is the double one, first of harmonising them by assigning to them their right relative status as abstractions, and secondly of completing them by direct comparison with more concrete intuitions of the universe, and thereby promoting the formation of more complete schemes of thought.... Philosophy is not one among the sciences with its own little scheme of abstractions which it works away at perfecting and improving. It is the survey of sciences, with the special objects of their harmony, and of their completion.

Now this suggests that the primary objects of philosophic study are the abstractions and perhaps also the methods and results of the special sciences, and that it aims at some kind of synthesis and completion of the sciences. It seems to me that this is, on the one hand, to assume an impossible burden, and, on the other, to restrict unduly the scope of philosophy. An impossible burden for the reason that we are asked to unify an indefinite plurality of sciences, no one of which is a unified whole in the first place, or has any prospect or need of becoming so. An undue restriction, because all the sciences together are only one set of institutions, and philosophy has no reason to confine itself to this set and ignore others, or even to give pre-eminence to this set. So I propose to describe philosophy as the critic, not of abstractions, but of institutions in general, of which the sciences and their abstractions are a quite special kind.

By institution, as a first approximation, I shall mean any provision or arrangement of means or conditions for subsequent activity, additional to or in modification of the means or conditions that are already present prior to the institution, whether present in nature prior to all institution or present in nature only as modified by previous institutions.

And now I must crave your particular indulgence. In order to develop the notion of institution and the notion of criticism sufficiently for the purpose in hand, I must say some things that I cannot take time to justify, and some which I am not sure I could justify in any length of time. It is of course a retiring president's privilege to commit all the fallacies that have been exposed in the working sessions earlier in the day. Until further notice, I shall be reading as statements what are for the most part really queries, and I ask you therefore to keep supplying "Is it not true that...? ", or "May it not be the case that...? ", or "Will you permit me to suppose that...? ". It is only to save time and to avoid the monotony of so long a series of questions, that I shall use the declarative form. I believe that, after making the corrections you would wish me to make, and inserting all the qualifications which a close scrutiny would find to be required, I would have enough left to warrant my later proposals. But that also I must submit to your judgment.

The notion of institution, in a closer approximation, involves those of purpose and choice, will and decision; that is, of the arbitrary. It is the notion of what would or might have been otherwise if the purpose had been different, but of what might also have been otherwise to the same purpose. It is the notion of what is subject to criticism in the light of the original purpose if that can be found, or in the light of any purpose that may have taken its place; and of what is alterable by subsequent decisions, but never so alterable as to cease to be arbitrary.[1]

In a general way, the purpose is to supply the means for, or to give scope, opportunity, protection, assurance, effectiveness, direction, form or style to future activities of a certain description; or to prevent or discourage activities of other descriptions. When agriculture, industry and commerce have been instituted, we can work and we can supply ourselves with the necessaries and commodities of life. When games have been instituted, we can play. When the arts have been instituted, we can create and enjoy objects of beauty. When religion has been instituted, we can worship and pray and give solemnity to the great occasions of life. When schools have been instituted, we can learn and we can learn how to go on learning. When laboratories and libraries and museums and observatories and sciences have been instituted, we can engage in research. When government has been instituted, we can give some sort of working harmony to the other institutions, which we may call their

[1] By "arbitrary" I mean not "unreasonable" but "dependent on will"; more exactly, I mean "residually arbitrary" in the sense that, when reason has done what it can, discretion remains and commitment is still required. The objectivity of value is not thereby impaired.

constitution, or we can stabilize and protect a constitution which they already have, and adjudicate the conflicts that threaten it. These and all other institutions depend upon the two basic ones of family and speech. The length of human infancy makes obvious the primacy of the family. When instituted signs have been added to natural signs, we can speak and write, read and think; when the arts of rhetoric and literature, and learned societies and occasions for ceremonial speech have been instituted—we can make and suffer presidential addresses.

Every value is conditioned in one way or another by institutions, and all valuing tends to take on institutional form, to strengthen or to weaken or otherwise to modify existing institutions, or to give rise to new ones. To paraphrase Aristotle, the basic institutions come into being in order that men may live, and they continue in being, and others are added, in order that men may live well. The institutions of a society, in their order or constitution, are the conditions and means of those activities in which the good life consists in that society. They are also the matrix out of which the conception of the good life is developed. Every institution has its particular value or values, and develops its own rationale, and perhaps even a distinctive type of rationality. Out of the conflicts of institutions, the strains upon individuals that result from them, and the efforts at adjudication and reconciliation, there develops a general theory of the good life, and a generalized rationality, which is instituted as the continuing critic both of other institutions and of itself. It constructs the theory in terms of which the institutions of a society are justified to itself and to its neighbors, in terms of which internal conflicts are adjusted and dissatisfactions quieted, but also in terms of which the existing institutions are weighed and found wanting, and alternatives are conceived, advocated, and instituted. This continuing critic is philosophy.

Already in the eighteenth century Vico had derived the logic, metaphysics and ethics of Socrates, Plato and Aristotle from the disputations of the market-place, the public assembly and the law courts of Athens. We have since learned to suspect behind the earliest philosophical speculations in the Greek world the desire to make or to resist innovations in the institutions of the city states of Asia Minor and of Southern Italy and Sicily. We shall not much exaggerate the prevalent view if we say that philosophy began not in cosmic but in civic wonder; that its cosmogonies and cosmologies were politically inspired and had political applications; and that it conceived the world order in the image of the order of institutions in society before it began to use the former as a standard by which to criticize the latter.

In any case, the critic of institutions needed a metaphysics as well as a general theory of value or of the good life. Eventually it needed all the philosophic disciplines so far instituted, and it will need others still to come. But philosophy has moved during most of its history between the poles of politics on the one hand and metaphysics on the other; between the life of reason and the realms of being. I have no doubt that it will continue to do so.

The most general distinction in the history of philosophy has been that between nature and institution, and the basic problems of metaphysics turn on the relation between the two. Most of the other recurring distinctions are variations on this one, such as nature and culture, fact and value, matter and form, particular and universal, body and mind. Whatever we mean by mind, our best approach to it is not by introspection but through institutions. But though most of the concepts and distinctions of philosophy have been developed out of the theory of institutions, we forget their origin unless a traditional label reminds us, as in the distinctions between natural and instituted signs, law by nature and law by institution, or law by divine institution and law by human institution. We subject our concepts and distinctions to dialectical refinements apart from the theory of institutions, and we seldom subject them to the final test of being brought back home.

Let me illustrate the institutional setting of the problem of universals. An institution that is to survive must be so set up and set going as to become a unit in a system which includes other units of the same kind, and includes other kinds as well. A family must take its place in the family structure of society, in which it sustains certain relations to the families of groom and bride, which reproduce themselves in and through it, and other relations to other families in the neighborhood, the church, the P.T.A. A bank must find its place in the banking system, a school in the school system.

A new university, for example, will conform in most respects to a university type or pattern already exemplified in other universities. But this need not mean merely that its founders are fashioning their university after the idea of a university, whether Newman's or Plato's or any other. Peirce, for instance, defined a university for the Century Dictionary as a research institution. The editors asked if it was not a teaching institution. Peirce replied that it was not and never had been, and that we would have no universities in this country until we learned better. But it is less the attractiveness of Peirce's idea than the demands of government, industry, business, and agriculture that are gradually producing an overwhelming preponderance of the research over the teaching function in our universities. Meanwhile, a university has to fit into a system of higher education, to receive students who are graduated from high schools and transferred from other universities and colleges, to exchange credits, to have its graduates accepted on an equality with those of other universities. The idea may prevail in the long run, by repeated appeal to it as a standard of criticism, but in the short run the authoritative universal is less the Platonic idea, less the eternal object or the ideal limit or the abstract universal, and more the Hegelian concrete universal, the institutional system.

Was there a first institution? If so, it will not be the historian, the archaeologist or the anthropologist who can take us back to it. Was there a first family, a first church, a first school, a first bank? If not, we have only to extend the list, and we are sure to come to an institution of which there was a first, but it will not have been the first institution. Every act of instituting presupposes an existing set or matrix of institutions. We make use of old institutions in creating a new one, and the new must have from the start some continuities with the old, and is bound to acquire others, until its roots are sunk in most or all of them. Institutions have institutions for their parts, and are parts of institutions; institutions become reciprocally parts of each other; and there is no all-embracing institution which has all others for its parts but is no part of them.

Universalia ante rem, *in re*, and *post rem*. If I have not quite clearly illustrated all three, it is apparent that it would not be impossible to do so.

Or try matter and form, body and mind. Institutions cannot be pointed to in the sense in which we can point to physical objects or to human beings, or to certain of their details or qualities or movements. We cannot even point to parts of institutions, because the parts of institutions are also institutions. But there would be no institutions without human beings, and physical conditions, and perhaps there are none without some embodiment in physical objects as well. In some institutions the material embodiment is so prominent that in moments of absentmindedness we may come close to identifying the institution with it. The supermarket then becomes a stock of goods, the countryclub a golf course, the court a court house, the utility corporation becomes the utility, the church a building with altar and pews, and similarly with banks, schools, and research institutions. That is, we tend to identify what cannot be pointed out with what can. This is not always an unfortunate tendency. To redesign the material embodiment is the most obvious and often the most effective way of changing the institution. If we wish to have discussion instead of lectures in certain courses, the first step may well be not a directive to the teacher but an operation on the classroom, taking out the fixed seats that face forward, and putting in a large oval table with moveable chairs around it. That may suffice without the directive. We change our minds, or may change them, by changing the circumstances in which we act.

Yet the institutions cannot be analyzed into the human beings and the physical objects, or even into the behavior of the former and the shapes and movements of the latter. They are not sensible, but intelligible forms; forms we cannot perceive, but cannot think away without thinking the institutions away; forms, indeed, without which we can scarcely think at all.

We cannot adequately criticize institutions without conceiving alternatives. For purposes of criticism, the scope of an institution includes that to which, taken as a whole, an imagined alternative whole is preferred which is like the given whole in some respects and unlike it in others. Institutions are in this sense relative to our powers of comprehension. But our powers of comprehension have developed by experience of the way in which, within a whole, change at one point brings change at others, so that the question of the desirability of a whole which includes all the changes, as compared with a given whole in which none of these changes has as yet taken place. Furthermore, comprehension is not ordinarily immediate or direct. We do not grasp an institution as we do a visual work of art which may be taken in as a whole from any of several points of view. Institutions are comprehended by the help of maps and charts and models and complex theoretical descriptions and historical narratives. The techniques of comprehension may become so specialized to particular institutions or sets or types of aspects of institutions, and so focused on description without reference to the purposes of criticism, that an institutional or social science detaches itself from the critic of institutions. All the present social sciences have in this way detached themselves in the nineteenth and twentieth centuries, as the natural sciences already had in the seventeenth and eighteenth. And the critic is left without command of the apparatus and techniques of comprehension.

Interlude

As an interlude at this point, I should like to read the scenario for a three-act skit called "The Progress of Philosophy."

Act. I. The world is back stage and back drop. The philosophers are studying and criticizing it. After prolonged study of a part of it, one of them addresses the others, and they stop and listen. As they listen, they do not look at him, but at the part of the world he is talking about. Now and then they interrupt, point at something, and correct him. This is repeated for other parts of the world. All the philosophers move freely from part to part, and there is not much deference to the authority of those who have lingered longest over a particular part. But off to the right a group of them are constructing some very complicated apparatus and building a fence around it and bringing in pieces of the world through a gate. They have stopped calling themselves philosophers and they paint a sign on the fence, saying, in big letters, PHYSICS, and in smaller letters underneath, "Philosophers not wanted."

Act II. The world is completely blocked off from view by a row of inclosures, with such further labels as CHEMISTRY, BIOLOGY, PSYCHOLOGY, ECONOMICS, SOCIOLOGY. A few philosophers wander back and forth complaining about the inclosures and recalling the good old days when this was known as "The Philosophers' Common." One of them gives another a boost to look over an inclosure at the world beyond. The philosopher on top reports that there are gates on the farther side through which the scientists go out into the world and return. Meanwhile other philosophers have stationed themselves at knotholes in the fences. Now and then they take peeps to see what the scientists are doing, but for the most part they hold their ears to the holes and listen to what the scientists say, and exchange reports with each other, and critical comments. One philosopher suddenly shouts through his knothole, "Oh! Mind your language!" A few moments later he whispers, "Would you like to have me mind it for you?" The philosophers form a chorus and sing the philosophers' theme song. When they come to the refrain, each of them approaches his knothole. The refrain is, "Mind your language! Or would you like to have us mind it for you?"

Act. III. We hear more bustle than ever in the inclosures, and the banging of the gates suggests much traffic with the world. But the philosophers have not only forgotten the world but have turned their backs on the sciences and are sitting in two huddles down front. Those at the left are playing a game of cards which is supposed to settle the pecking order in philosophy. As they lay down their last cards, one says, "Object!"; the second, "Meta-!"; the third, "Meta-meta-!"; and the fourth, "Meta-meta-meta-!" But before proceeding to peck, they are distracted by a speech which has begun in the huddle at the right, and they straggle over to hear it. The speech is to the following effect. "Since the scientists won't let us mind the world, or *them*, and they don't much care whether their *language* is minded or not, and we can't mind it very well through the knotholes anyway, why not start minding *ordinary* language? That's something they can't deprive us of, and we always have it with us, so we needn't budge from where we are." After some discussion, they decide that that is what they should have been doing all the while, so that what seemed like a forced retreat from the world and from science was really an advance to the rear, and this is what progress in philosophy consists in. So they take out notebooks and start scribbling, and then they read each other short papers on such subjects as, "Am I now dreaming? " "Can I feel your pain? " "Can we witness or observe what goes on 'in our heads'? " "Can a man witness his own funeral? " At the end they stand and sing their second song: "Doesn't it strike you / And greatly surprise you / That ordinary language / Is so very extr'ordin'ry? " (*Curtain.*)

Now if my conception of philosophy as the critic of institutions is sound, philosophy must stand in a relation to the social sciences very different from that in which it stands to the natural sciences. Philosophy needs the social sciences, and they need it, a great deal more than it needs the natural sciences, or they it. For philosophy, the natural sciences are primarily a set of institutions, institutions of research and teaching, intimately linked with agriculture, engineering, industry, medicine and other technological institutions. Philosophy, as the critic of institutions, is concerned with all of these primarily as objects of value; that is, with reference to the activities to which they give form and direction, and to the value of those activities in the good life as a whole. The objects of the natural sciences themselves are not the philosopher's primary objects. They belong to that side of the nature-institution couplement which concerns him only secondarily. He needs to revise his metaphysics from time to time in the light of general developments in the natural sciences, but he seldom needs a detailed knowledge of their findings. On the other hand, while philosophers of science have doubtless contributed in the past to the understanding the natural scientists have of their own procedures, and have perhaps even contributed some refinements to the procedures themselves, these are now well established and it does not seem probable that in future the natural scientists will greatly need philosophy in their own proper business, though certainly they will need it whenever they attempt to turn science into metaphysics or to erect a metaphysics on the conclusions of science, with an imperfect realization of the institutional character of science, and of the necessity of bringing institutions other than scientific into the metaphysical scheme.

With the social sciences the case is quite different. We cannot content ourselves with considering them as institutions among others, for their objects are our objects too in a way in which those of the natural sciences are not. Our chief purpose here must be not to study the social sciences but to study, with their help, what they study. Every institution, including language, is the concern of one or more social sciences, and the social sciences are continually making discoveries in ignorance of which we cannot continue or resume our traditional function. If we resign ourselves to ignorance of the social sciences, we must resign ourselves to being critics of only one institution, namely language, and very ignorant critics of that.

On the other hand, the social scientists need us. There is great confusion among them as to their functions, methods, and concepts. Some of them, aspiring to the condition of natural science, would turn the social sciences into behavioral sciences, values into observable natural facts or events, and institutions into behavioral patterns. Yet these same scientists may be employed as experts to evaluate alternative institutional arrangements, and may irresponsibly assume the values of those who employ them, or assert their own values, without any attempt at the objectivity they profess in their research. Other social scientists are convinced that values are of the essence of what they study, but that neither institutions nor values are data of the natural science kind. They would like to deal with them scientifically, but they do not know how. Finally, there are social scientists who, often without knowing it, are really philosophers, interpreters and critics of institutions, who view the industrious value-free researches of their colleagues with a disdain which few professed philosophers would permit themselves. Now, as Peirce said to Royce, "I wish you would study logic—you need it so much," so we might fairly say to all three groups of social scientists, "We wish you would study philosophy—you need it so much." In all three groups, however, and in the many intermediate groups as well, there is already a minority of social scientists who have some sense of the relevance to philosophy of what they are doing, and who are willing to believe that philosophy might help them toward a more rational ordering of their own endeavors. They would welcome us to their counsels, and would be glad to be welcomed to ours. They would accept our services as friendly critics, and even invite our collaboration.

I have now finished the part of my paper for which I begged your particular indulgence, and I come to my proposals. Let me repeat that they do not include a proposal that any of you should practice philosophy in any other way than that in which you are now practicing it. I believe that you are all philosophers as I understand philosophy, and that, if it does not seem so to you, that must be because I have failed to make myself clear. I do think, however, that a series of developments in academic history has led to concentration on the institution of language, and comparative neglect of many others; and I think that another development in academic history now offers us an opportunity to redress the balance. That new development is the great increase in college and university enrolments which is already well begun. I base my proposals on the premise that, within a future for which we may reasonably plan, there may be twice as many persons making a career of philosophy as there are now, and that the intellectual and cultural vitality of philosophy may be recovered by a distribution of effort among those who join our ranks different from that which now obtains among us. I have two sets of proposals.

The first set is addressed to all of us as advisers of individual students. It is that we be on the alert for students who seem to have some aptitude for one or another of the social sciences, and who have not already given hostages to fortune, and that we urge them, after a thorough grounding in the more technical philosophic disciplines, to obtain a similar discipline in a social science of their choice. They should carry their studies far enough to have had first-hand experience of the research methods of the science in question, and, if possible, be admitted to membership in the profession. They should subscribe to at least one representative journal, and cultivate the friendship of at least one outstanding young scientist in the field, through whom they may hope to be kept informed of current developments as from within the science. They should attend its departmental seminars and professional meetings, and in due time offer papers of their own. They should turn their growing scientific knowledge to philosophic account, and make it available to their fellow philosophers so far as that is desired.

For purposes of such advice, I think we should include history among the social sciences, and I think we should encourage some students to turn instead to law, public administration, business management, labor union organization, social work, and even to supplement professional training by practical experience in these fields.

My second set of proposals is addressed to the philosophy departments in our larger universities. Without diminishing, and perhaps while even increasing, the absolute numbers of persons with strength in mathematics, natural science, languages and literature, and the other arts, I suggest that they seek to increase greatly their relative strength in the social sciences. If my conception of philosophy is sound, a department with twelve members, for example, might reasonably have an expert in each of the major social sciences and in law. These men should be encouraged to spend part of their time in cultivating their several sciences, and in maintaining good communications with the respective departments. We should not expect them to begin publishing as soon, or to publish as often, as their colleagues in logic, the philosophy of the natural sciences, semantics, or analysis. They would of course be writing, and they would present papers in our meetings as well as in social science meetings. Many of these would never be published, or would be published only after many metamorphoses. If philosophy becomes again the general critic of institutions, the expertness required for useful publication will demand a greater maturity and a wider experience of life, in addition to the logical, analytical, and dialectical skills we already prize.

If there are really able young men and women who are willing to extend their preparation for philosophy in the way I propose, and if college and university departments of philosophy are ready to add persons so trained to their staffs, and promote them on evidence other than print, I do not doubt that the foundations will supply the necessary scholarship funds for a trial period, until the universities themselves can make suitable provision.

We might also ask the foundations to support for several years a summer session in which leaders in the social sciences would conduct seminars at an advanced level for philosophers. I think the foundations can be persuaded that philosophers are worth educating in this and other ways. For example, another project that would require foundation support would be the preparation of a classified bibliography of social science literature carefully selected and critically annotated for the particular purposes of the general theory and critic of institutions. The continuation of such a bibliography in periodic supplements might then be undertaken by one of our university departments which had expanded its staff to include a corps of experts in the social sciences. Such a department might also enlist foundation help in launching a journal which, along with *Ethics*, would lead the way in the kind of recovery of philosophy toward which all my proposals are directed.

Two concluding words.

1. It may turn out in fifty or seventy-five years that the social sciences have been a mistake. That is, it may turn out that their effort to approximate the mind-free value-free concepts and methods of the natural sciences, and their search for generalizations of the natural science kind, have been a mistake. The social sciences may then return to the condition of philosophy, each functioning as critic of institutions from the base of a particular institution or set of institutions. This would be unobjectionable, and it may perhaps be necessary. All institutions are implied by any one, and working out as it were from a particular one may yield as adequate an understanding as attempting to survey them all without focusing sharply on any. More exactly, a just survey is possible only after working out from each to the others.

Meanwhile, however, we must live with the social sciences as they are, and on their terms as well as ours. If we do not, they will be driven by the pressure of rapid social changes to create their own value-orientation, their own synthesis, their own general theory and critic of institutions. If God did not exist, it would be necessary to invent Him; and if the critic of institutions no longer lives, and cannot be resurrected, in departments of philosophy, it will be necessary to create it outside of them.

2. Fortunately for most of us here tonight, philosophy has other resources besides the social sciences and besides its own more technical disciplines. It draws nourishment from all the humanities as well, and in the end it performs its critical function in the humanistic rather than in any scientific way. Thus the history of philosophy has an importance which the history of science cannot have. The history of philosophy is philosophy itself taking its time, and its way of taking its time includes not merely a continual bringing forth of things new, but a continual review of the old. It continually re-sifts, re-selects, and re-orders its past creations, re-edits, re-translates, re-reads, re-interprets, and criticizes afresh. Its great classics do not diminish but grow in power. The art of teaching is itself often the art of bringing the thoughts of our students back again and again to a sentence of one or another of the great philosophers, until our students grasp the significance which only the entire history of philosophy before and since packs into that sentence, and until they bring to bear upon it all the critical resources which that history affords. Thus the institution which is eminently the critic of all others is also that which, more than any other, is critical of itself.

The Philosophy of History: A Dialogue*

by Max H. Fisch

(*Offprinted from *Philosophy* (Tetsugaku) 36: 149-170, 1959—published by Mita Philosophy Society, Keio University, Tokyo, Japan)

Persons of the Dialogue: a Philosopher and a Historian
Time: April 14, 1957
Scene: the Historian's study

P. I've been reading lately what the philosophers have been writing about history. They are having a symposium about it at Chicago in May, and I've been asked to lead off with something for my fellow symposiasts, Donagan and Hinshaw, to shoot at.(1)

H. Oh? I thought the philosophers had long since lost interest in anything so vast, windy and interminable as history. Has the pendulum swung, and are they beginning to take large views again?

P. No, it is not large views of the human past they are taking, but minute views of the historian.

H. *The* historian? A fictitious character, I presume?

P. Now don't be captious; leave that to us. I admit they sometimes seem to be imagining themselves doing history, and to be analyzing what they imagine themselves doing. But sometimes they examine short passages from real historians.

H. Dead ones, you mean? It would spoil the fun if the historians were alive to answer their questions for them. And what kinds of questions do they raise?

P. They are nearly all of one kind, epistemological. What is the historian's object, how does he come to know it, and in what does his knowledge consist? Are his statements verifiable, and, if so, how? Is moral judgment a part of his business, and, if so, how does he reach such judgments, and are they confirmable? Can he be objective in his statements and judgments alike, or in his statements only, or in neither? Does he treat his dated and placed events as unique, and as material for his own narrative only, or as susceptible of generalization also; and, if the latter, does he prepare them for the uses of the generalizing social sciences, or do his own generalizing? Does he explain the events, actions, thoughts, or other changes which he reports or narrates? Does the Popper-Hempel model fit all his explanations, or some only, or none? Are other models feasible, and are they required? (2)

H. The questions are fine, but how do the philosophers hope to answer them?

P. For the most part, by watching closely what the historian does, though I won't deny they proceed sometimes as if the power of definition and a certain dexterity in linguistic analysis would do the trick.

H. If the questions are honestly inductive, and inductively honest, I should think a wide induction would be in order. There are many historians, and they do many different things. Has a sampling technique been devised? Is there a repertory of samples? Are historians invited to submit additional samples, to represent kinds of historical performance they think are not so far represented? May it not happen otherwise that explanation, for instance, is defined without regard to what historians do, and it is then decided that whatever they do that is not that, is not explanation, or is at best an explanation sketch?

P. You have taken the words out of Hempel's mouth.(3)

H. Why, in the work even of a single historian, taken almost at random, I wager more different things would be found than are dreamed of by your philosophers.

P. That is a hypothesis worth trying. If I knew a historian of modest parts, old enough to have done every kind of thing he is likely ever to do, but lazy and unproductive enough to permit my examining the whole of his published work, I might make the trial myself.

H. Here am I; try *me.* I've been writing history for thirty years. You've known me all that time. You claim to have read everything I have written. Besides, it is all here on my shelves or in my files, and we can refresh our memories. Will I do?

P. Hm... At least for a start. As a historian, you're an ill-favored thing, but mine own. 'Tis a poor humor of mine, sir, to take that that no man else will. If I may be your Touchstone, what rubs onto me from your work shall be my philosophy of history, until Donagan and Hinshaw teach me better.

H. Where shall we begin?

P. Where *you* began. How did you find your way into history in the first place?

H. I was brought up on Biblical history and Biblical criticism, higher and lower, J and E, first and second Isaiah, the prose folk tale of Job and the dramatic poem sandwiched into it, and all the rest. As a child, with scissors and paste, I made my own harmony of the gospels.

P. Well, that qualifies you as an amateur in scissors-and-paste history, which Collingwood says was nearly all there was until recently, and in critical history, which he says was the last phase of scissors-and-paste history. The question is whether you ever graduated into what he calls scientific history.(4) Let's come down to your prentice job, your thesis on "The Influence of Stoicism on Roman Law."(5) Let me see again that carbon copy on which I once made so many marginal comments. I remember our having much argument as to what might constitute influence, and what might be evidence of it.

H. Here it is, and right you are.

P. You must have been under the influence of something heady when you took on such an impossible thesis. What was it?

H. Oh, they had taught me in college that besides Biblical history and literature the chief ancient sources of the intellectual substance of western civilization were Greek philosophy and Roman law. Then, in graduate school, in the scholarly literature I came upon statements to the effect that the philosophy had had this or that influence on the law, and, through the law as well as directly, on the Christian theology whose other chief source was the

Bible. I knew the Bible, and I had studied the philosophy, but not the law. I had to write *some* thesis. I foresaw no other occasion for studying the law, so I took that one, and nobody had the pity to stop me.

P. The title of your thesis would suggest to our philosophers that certain features of Roman law are to be explained as effects of certain features of the Stoic philosophy. Your table of contents would suggest to them that you are trying to show how those causes came to have those effects. You do this by exhibiting the intermediate links. You trace the Roman reception of Stoicism, and the emergence of Stoic ideas in Roman religion, education, literature, and science, and in the commonplaces of moral counsel and public speech. You then trace the same ideas and other more technical Stoic doctrines in the detail of Roman law. As I recall, you found more Stoicism there than any previous historian had found, and you went farther toward explaining its presence there. Now would you agree that the influence you assert is a form of causation, and that causation is intelligible only in terms of causal laws? If so, the standard model for what you were doing would seem to be the Popper-Hempel or covering-law model.

H. I can't think what the covering law in this case would be. That in any society in which a certain philosophy prevails in other aspects of culture it will prevail in law also? But I nowhere assert that, and I nowhere assume it.

P. Then how does the prevalence of Stoicism elsewhere explain its prevalence in law?

H. It certainly doesn't if an explanation must logically entail what it explains, and if general laws are required for that purpose. Unfortunately all the universal propositions available to the historian for conferring necessity upon his conclusions are propositions he knows to be false. And he suspects that if he took one of them and quantified and qualified it until there were no longer any exceptions to it, there would also be no instances of it but the one he set out to explain.

P. But if you forego that sort of explanation, what have you left to offer?

H. In the present case, I merely assume that my reader, when he has before him the evidence that some of the lawyers were Stoics and all of them lived in a society in which Stoic ideas were current, will agree with me that it would not have been unnatural for them to do some of their legal thinking in Stoic terms, so that what looks like Stoicism in the legal texts may really be Stoicism. On this assumption I proceed to explain the texts by placing the doctrines they contain in the framework of the Stoic system. Surely explication of texts is something all historians do, and surely it is not any kind of causal explanation, and calls for a model or models quite different from the covering-law model.

P. I agree that if it were simply a matter of your telling a story based on evidence, and of the reader, believing your story, being thereby prepared to understand the texts as you explain them, the Popper-Hempel model would not fit. But it still seems to me that you are doing more than that. For example, you explain the reception of Stoicism at Rome in part by its congeniality to the Roman character, and I see I wrote this comment into your margin: "Perhaps there has never been another people with such a taste for edification as the Romans, and there was no Greek philosophy so ready as Stoicism to do the edifying."

H. The coverg law in that case would presumably be that character determines philosophy, or determines a choice among the philosophies offered. But characters are habits or dispositions, and surely dispositions are not causes in the covering-law sense.

P. Objection sustained. It just happens that Donagan, starting from Ryle's distinction between law statements and dispositional or law-like statements, is in a fair way to work out a model of historical explanation in terms of disposition statements. But it appears that explanations on the Ryle-Donagan as well as on the Popper-Hempel model will logically entail what they explain.(6)

H. What a lust for necessity you philosophers have! Such knowledge is too wonderful for me; it is high, I cannot attain unto it. I should have said my explanation amounted to no more than this. Choosing Stoicism was of a piece with other Roman choices and actions; there is nothing surprising about it; no special explanation is required. Or, if you prefer, whatever is to explain that must first explain the Romans; which I leave to those who are better equipped.

P. Let's move on, then, to your later works. But first, why did you never publish the thesis?

H. It was immature and incomplete. I needed to take more account of philosophies other than Stoicism, and I needed a wider and firmer grasp of Roman law. There was great controversy as to the extent to which the surviving legal texts had suffered from interpolation, and many of my texts were suspect. I spent a decade revising and reworking. Meanwhile, during the Great Depression, apart from rude shocks to the foundations of the society in which we lived, I encountered two shocks to the foundations of my thesis, and I have still to finish the repairs. The first shock was administered by the historian Tarn, and the second by the philosopher Vico.

P. Now don't forget Marx. When so many of our friends were turning Marxist in theory, if not Communist in party, I remember your reading him extensively and closely for what I believe was the first time, and giving a more sympathetic account of him than I imagine you would now.

H. So I did. The point that came home to me was the way in which law and philosophy alike, along with the other non-economic aspects of culture, were alleged to be ideological reflections of the class struggle and of its dialectic. But, exciting as Marx was, and great as the pressure then was to master and assimilate what he had to say, he was for me a way-station to Vico. Marx had recommended Vico's *New Science* to Lassalle "for its philosophic conception of the spirit of Roman law in opposition to the legal Philistines," and it dawned upon me that I might be one of the Philistines. Instead of deriving Roman law in part from Greek philosophy, or deriving both entirely from economy, Vico derived Greek philosophy from Greek law; or, more exactly, he derived the logic, metaphysics, and ethics of Socrates, Plato and Aristotle from the disputations of the Athenian assembly and courts: first, popular government, he said; then laws; then philosophy. Roman law, like Greek law, was an indigenous growth, and the borrowings I had made so much of, if real at all, were late and insubstantial. So it appeared that the interpolation detectives were but following in Vico's footsteps at an interval of two centuries.(7)

P. Shock number two. And number one was Tarn?

H. Yes. He was the historian on whom I had most relied for my general conception of the primacy of the Stoa in Hellenistic civilization. Three years after my thesis was accepted, he gave a famous lecture before thyBritish Academy in which he removed all the props on which I had built.(8) Until then, he had presented the Stoics as heirs of Alexander, who systematized and propagated the ideas that Alexander had projected or inspired. Now Tarn disinherited the Stoics, and robbed them of most of the influence he had previously ascribed to them. Alexander was in fact "the first man known to us who contemplated the brotherhood of man or the unity of mankind." The attribution of those ideas to Alexander could not be a projection backwards of Stoic thought, because there was an "irreconcilable opposition" between Stoicism and those ideas, and the Stoics did not adopt them until two centuries or more later, when other philosophers had made them commonplace.

P. I seem to recall that you not only spent two or three years constructing an argument to show that Tarn had been right before and was wrong in his Alexander lecture, but you undertook to give a psychological explanation of his unacknowledged change of mind. And I think I shall have you there. May I see the article? Yes, here it is.(9) You say that Tarn had step by step been asserting his independence of Kaerst, the German historian of the Hellenistic age, and that he uncritically seized upon a specious essay by Goodenough on "The Political Philosophy of Hellenistic Kingship" to complete his emancipation. Were you not here relying on a psychological law which you did not explicitly formulate, and have we not here a clear case of the covering-law model?

H. Let's see. I show that in relation to Kaerst the Alexander lecture is but another step in a direction in which Tarn has been steadily moving, but that in relation to his own previous views concerning Stoicism the lecture is a nearly complete reversal. There is no new evidence to explain the change; there is only Goodenough's construction, from familiar but dubious evidence, of a non-Stoic political theory which Goodenough presents as the official philosophy of the Hellenistic age. Tarn dethrones Stoicism and puts this pretender in its place. If he is right now, he has been propagating for twenty years a radical error. He does not acknowledge his previous error, or his conversion from it. It seems clear that he is not aware of the extent of the change, or of the insufficiency of his reason for it. If he had been aware, he would have faltered, and he would not have been able to take that final step away from Kaerst which he wished to take. When an eminent historian grasps at straws and gives the lie unawares to much that he has hitherto stood for, an explanation is in order. I believe it is possible for a strong desire to have such an effect. For lack of a better explanation, I hazard a guess that this has happened in Tarn's case. I label it a guess; I say that even if it truly explains Tarn's latest views, it cannot explain them away; and I proceed to consider these latest views on their merits without further reference to my explanatory guess. Had I thought I needed a law to sanction my guess, I would have skipped it.

P. Are you sure you didn't suppress the law because it would have served equally well to explain your attack on Tarn by your need to cling to the primacy of the Stoa as a prime support of your thesis?

H. Touché. A hit, but not a wound.

p. Well, to get on: I don't suppose your various essays on the history of American philosophy have anything to do with your thesis?

H. Yes, they do, in this way. Marx and Vico had opened a view of the relations between philosophy and law quite different from that which I had assumed in the thesis. Now the fashionable theory of law in our society was legal realism. It seemed to stand in some logical relation to pragmatism. Was there also a historical relation between the two? Did either develop out of the other? If so, was the movement from philosophy to law, or from law to philosophy? This was the primary focus of all my researches into the origins of pragmatism. The fact that Holmes and Green and Warner, three of the founders of pragmatism, were lawyers, seemed to me significant,(10) the more so since Peirce had called Green "the grandfather of pragmatism." I traced legal realism and the prediction theory of law back in Holmes's published writings to 1872,(11) and verified Peirce's claim, under Green's inspiration, to have drawn pragmatism as a corollary from Bain's theory of belief, but at a time when the legal theory was already in print.(12)

P. In thus moving from the relation between Greek philosophy and Roman law to that between Greek philosophy and Greek law, and finally to that between American philosophy and American law, were you not preparing the way for some sort of comparison, and for a hypothetical generalization which might then be tested by study of the relations between philosophy and law in other societies?

H. Not at all. I had been interested in Greek philosophy and Roman law, not as samples of philosophy and law in general, but as the particular systems out of which western intellectual culture had developed. I was interested in American philosophy and American law as the particular systems of our own more immediate society. To be sure, what I found in the American case suggested questions concerning the Greco-Roman case, and vice versa. I was struck by a certain kinship between Stoicism and Pragmatism, which had been a source of embarrassment to Peirce, but this was nothing to my purpose. I was not in search of generalizations of that kind. I did eventually try my hand at another kind of generalization, much more common among historians, for which the philosophers have not yet given us a model.

P. What was that?

H. A general characterization of what I called "The Classic Period of American Philosophy," the period from the end of our Civil War to the beginning of the Second World War. I presented this in two slightly different forms, one an unfocussed statement,(13) the other a statement with Dewey as focal figure.(14) In both forms I generalized in terms of the major influences in the prevailing climate of opinion, and of the major themes and tendencies common even to philosophers the most critical of each other, and distinguishing them all from those of other times and places. Each of my generalizations was illustrated by quotations from all or most of the major philosophers of the period, each quotation being at the same time characteristic of its author.

P. If our philosophers were to provide a model for that sort of mutual or reciprocal explanation, it just might turn out to be a variation on Hegel's concrete universal, or identity in diversity, the identity affording an explanation of the diversity and of each diverse detail, and the diversity being an explication of the identity. Hegel is a bit out of fashion, but I see there's going to be a Hegel symposium following ours. Perhaps it will start something in this direction. Or perhaps a model not yet dreamed of will be found to fit.

In any case I am still puzzled about your continually shying away from the covering-law kind of explanation. What puzzles me is that Vico is one of the philosophers you have most admired and studied, and about whom you have written most. Now Vico was surely a philosopher of history, and he presented his philosophy of it under the title of "Principles of a New Science concerning the Common Nature of Nations." His ideal eternal history traversed in time by the histories of all nations looks very much like a general law. He at least wanted history to be a science. Don't *you* want to be scientific?

H. Yes, if it means doing better what I am already doing. No, if it means doing something different. Mind you, I had no objection to Vico's search for law, and I have none to other historians seeking it. If they succeed, I shall be instructed; if they continue to fail, they will take the curse off my laziness. I only say that I don't seek it, and that it is my impression from a wide range of reading in history that few historians do. And I confess that it was something of a problem for me why Vico had made so much of his ideal eternal history. That seemed to require explaining. I thought I found the explanation in the scientific movement at Naples in his youth, and particularly in a scientific academy called the Academy of the Investigators. Its most flourishing period was the five years preceding Vico's birth, but there were two revivals of it during his lifetime, and he was a member of it during the second revival if not also during the first. I thought it was the prestige of that Academy that led Vico, in emulation of the new sciences of nature which it cultivated, to construct a new science of history that should have the same generality or universality, and a higher certainty. After several years of preparation, I spent a year writing the history of the Academy and tracing the lines of its possible influence on Vico.(15)

P. Is there not a general law assumed in your thus explaining the genesis of Vico's new science?

H. A law to the effect that when historians live in a society in which natural science has high prestige and accepts the covering-law model of science, history will tend to do likewise? The only way to keep that from being falsified is to keep it vague; but, let it be precise or vague, I have no use for it. It seemed an hypothesis worth trying that Vico in particular was influenced by the Academy of the Investigators in particular, with no general law in the premises. If he was so influenced it should be possible to find fresh evidence of the influence. I found the evidence, confirmed my hypothesis, and thereby provided a partial explanation of Vico's excessive addiction to the covering-law model of history. So far as I was then or am now aware, I did not at the same time confirm any more general proposition than that.

P. I think the kind of generalization you practice, that of "the character of the age," may depend upon the kind you forego; but let that pass for the moment. It is time we made a first bow to the question of objectivity and subjectivity in history. Would you admit that your interest in the Academy for its bearing on Vico gave a certain subjective cast to your history of it?

H. I admit that I mention Vico's name more frequently, and put more emphasis on his relations with the Academy and with its individual members, than a historian with a different interest need have done. But I do not admit that this resulted in any misrepresentation or distortion of fact. Without that interest, I should never, after long preliminary study, have devoted a year to the Academy. I searched the libraries and archives of Italy for source materials and for accounts based on source materials no longer available. I made use of every scrap of evidence I could find, and I appended a complete bibliography in which I reported locations of manuscripts and rare publications, so that if anyone cares to check my statements, he can do so with a minimum of trouble. I think I have shown by example that a highly specialized interest is not incompatible with a full and balanced history, meeting the requirements of scholars who do not share that interest but have other equally specialized but quite different interests, so that no one of them need feel obliged to repeat any part of my labor.

P. May I return now to the question of generality? The Academy of the Investigators was one of many scientific academies. Surely you took account of that fact and considered it as a member of a class?

H. I took account of its relations, by correspondence and exchange of visits and by repetition of their experiments, with the other scientific academies of Italy and of western Europe at large. I ranked it next after the Lincei and the Cimento. I searched the archives of those and other academies, including the Royal Society of London, for evidence of relations with the Investigators. I offered summary statements about the rôle of academies in general, and of the Investigators in particular, at that particular juncture in the history of science. For example, about their relation to the universities in the Quarrel of Ancients and Moderns. Summaries of that kind have a large place in history, but they are nearly as time-and-space-bound as the particular statements that are directly warranted by the evidence. So far as I can see I neither asserted nor assumed a law.

P. From Stoicism and Roman law it seems a farther cry to the Academy of the Investigators than to Pragmatism and American law, but I see the bridge by way of Vico. But how about your other work in the history of science, and how about your work on the historian Binkley?

H. Vico had confirmed me in a belief I had already formed that the history of philosophy made insufficient sense apart from the history of science, and that medicine made a good focus for the history of science. (Vico, you know, had a theory of medicine on which he prided himself.) Well, under the name of intellectual history, I was teaching what was in fact history of science with special attention to medicine. I might nevertheless have remained an amateur in that field but for the series of accidents that put me in charge of the incunabula and the sixteenth and seventeenth century books in the Army Medical Library during the Second World War. Here was a chance to do something on the beginnings of modern science. Perhaps there is no better collection of source materials for that purpose, if medicine be taken as focus. The tercentenary of the first edition of Vesalius's *Fabrica* gave me in 1943 a sufficient occasion for trying my hand at a series of modest essays. What interested me was the relation between Renaissance science, Renaissance drawing and wood engraving, and Renaissance printing. I tried to trace the steps by which they were brought to a single focus in Vesalius's book.(16)

P. I am struck by the fact that, of all your writings, the essay on "Vesalius and His Book" comes nearest to being a piece of literary criticism, and I recall that the purple passage with which it concludes was used as a motto in Harvey Cushing's *Bio-Bibliography of Vesalius.* Is that sort of stuff the proper business of a historian?

H. Why not? The natural bias of the historian is toward individualization rather than generalization. He is not merely the recorder but the critic of the works of man. If he differs from the professional critic, it is in ranging more freely over the entire human past, and in working out in greater detail the steps by which the achievements of art and science became possible.

P. Your "Vesalius in English State Papers,"(17) on the other hand, is almost straight biography of the practicing court physician. Does any particular interest attach to it for our present purpose?

H. Not much beyond the fact that Vesalius's German biographer, Roth, though he had made use of the state papers and letters of the Empire, of Spain, and of several of the countries that maintained embassies at the two courts, had overlooked entirely the English state papers. It was merely historian's luck that from these fresh sources there emerged a vivid picture not only of the last twenty years of Vesalius's life but also of the institution of the court physician. At both courts, Vesalius spent most of his time as physician to the administrative and military staffs, to the ambassadors from abroad, and to the embassy staffs. He was in demand not only for his medical services but also for his company and for news of the emperor's or the king's health and of goings-on about the court.

P. In your book on Nicolaus Pol,(18) which I have more freshly in mind, you reconstruct the life of another Renaissance court physician who probably knew Vesalius's father and grandfather at court, and you reconstruct his remarkable library, and place his tract on the guaiac cure for syphilis in the history of syphilis. But from the point of view of the philosophy of history, the most interesting passages are those bearing on the controversy as to whether syphilis originated in the Old World or the New, or independently in both. From scattered bits of evidence, each slight in itself, you reconstruct the story of a medical commission sent by Cardinal Lang of Augsburg to Spain and Portugal to investigate and report on the guaiac treatment as it was practiced there by those who had learned it in Hispaniola. Will you hand me the book? .... Thank you....

Two things strike me here, and they are closely related.

The first is that the theory of the American origin of syphilis was one of the first conscious applications of the covering-law model to history. You quote the covering law as follows: "Such is the divine mercy that wherever it permits us to be afflicted for our sins, it places a remedy equal to our afflictions." By illegitimate conversion, this yielded the axiom, "Whence the remedy, thence the disease." The remedy, the wood of the guaiac tree, came from the New World. Therefore the disease for which it was specific, namely syphilis, came from the New World.

Now you don't subscribe to that covering law, but my second point is that you do seem to subscribe to others in the same context. Pursuing hints supplied by Paracelsus and by Ulrich Hutten, you proceed to connect both the medical commission of Cardinal Lang and the theory of the American origin of syphilis with the virtual monopoly in the guaiac trade enjoyed by those merchant princes the House of Fugger. "The prestige of the wood," you say, "was greatly enhanced by widespread acceptance of the doctrine of the American origin of the disease, and the spread of that doctrine was doubtless promoted, for the most part in good faith, by traders in the wood." Your covering law for that would perhaps be that doctrines are spread by those who have most to gain by their spread. You also suggest an explanation of the popularity of the guaiac treatment in terms of the class structure of society and of the medical profession itself. "The only really effective remedy then known," you say, "was mercury. It was commonly administered by inunction, which was a crude, untidy method suited for surgeons who treated the common people, but scarcely for learned physicians. The mercurial poisoning that followed was often more terrible than the disease itself. Only the harmless guaiac treatment, with its endless refinements and adaptations, gave full scope to the scholastically trained minds of the physicians. Only it was long and intricate enough, and called for sufficient variety of services, to gratify the vanity of their well-to-do patients."

H. I have three things to say in reply. In the first place, these are but parts of a total explanation. Among the other parts is the therapeutic fallacy. "Syphilis was not clearly differentiated from other venereal diseases; its selflimited character was not understood; the wood was credited with effects which would have followed from the diet and regimen alone, or even without their assistance. The patient took the decoction; the symptoms present when the treatment began disappeared in its course or shortly afterwards; therefore the decoction was the effective remedy."

In the second place, the trouble with the covering laws you read between my lines is that, as laws, they do not hold. I do not believe any of them, and I do not assume them, and I do not suggest that my facts lend them any confirmation. I merely say that there is some evidence that certain individuals were influenced in some degree by one or more of these considerations, and that if we extrapolate and assume that most or all of the believers in the efficacy of guaiac were influenced in varying degrees by these and by similar considerations, it will no longer surprise us that what we know to be false was so long and by so many as intelligent as ourselves believed to be true. The argument is not that people who have certain beliefs and certain motives will have certain other beliefs, but that it is possible for them, within the bounds of rationality, to hold such beliefs. Thus one may not unreasonably believe that guaiac regularly and normally cures syphilis (a) if it is not evident that it never does, (b) if it seems in some cases within a limited range of observation to do so, (c) if one believes that certain other things regularly happen which do in fact sometimes happen, and (d)if the guaiac cure can be subsumed under the formula for these other happenings. Now it sometimes happens that where a disease is common a remedy is also common which is in fact a specific for the disease in question. If one believes on religious grounds that this is regularly the case, and one is informed on good authority that syphilis is common where the guaiac tree is native and common, one is not thereby compelled to believe, but one may not unreasonably believe, that guaiac is specific for syphilis. I adopt this as a heuristic hint to guide my search for evidence. I find evidence that the people who believed in the guaiac cure did in fact believe one or another of the propositions under which that belief could reasonably be subsumed.

My third reply is that more important than the psychology or the logic of individual belief, in a case of this kind, is what we may call the vesting of belief in institutional structures, such as the merchant empire of the Fuggers or the complicated ritual of the medical profession. For example, the Fuggers had a model housing project at Augsburg called the Fuggerei, in which there was an infirmary named the Holzhaus or Woodhouse, where the guaiac cure was practiced with all its refinements, as a model to the entire profession.

P. I begin to suspect that before we are through you will have institutions doing the work my philosophers want covering laws to do. Let's get on. May I have another look at your edition of the papers of your historian friend Binkley? .... Thank you.... Your biographical essay, like the essays by him which you bring together, is full of ideas about historical research, writing and teaching, library and archival policy, the organization and reproduction of research materials. Perhaps the most interesting is the assumption that the objects of history are institutions, and the proposal to treat European history in such a way as to play down the institution of the national state and to play up other institutions, such as the family, the village and the city on the one hand, and on the other the institutions operative in what he called the world net of power, the world of debts and markets, and the world of opinion. Also to our present purpose is the fact that, like so many other American historians of his time, he was a relativist. In his view, as you put it, "every history, however remote the period with which it ostensibly deals, has its real terminus and controlling frame of reference in the time of its composition."(19)

H. Yes, and the two things went together in his case. He was working toward a history of Europe in the nineteenth and twentieth centuries which should be deliberately oriented toward the explanation of matters of urgent public concern at the time of composition and likely to continue to be matters of public concern for some years thereafter. He proposed to play down the national state and play up other institutions because the national state obsession was the chief obstacle to the development of the forms of order he thought were needed for our time.

P. Did you share his relativism, as I begin to see you did share his conception of institutions as the proper object of history?

H. I did not. Or rather I should say that the relativity he had in mind is a variable and admits of degrees. It is likely to be greatest in a general history that has its explicit terminus in the present, and may easily be much less in histories of remoter periods and of narrower scope, and least in those that admit of monographic completeness. I submit, for example, that my history of the Academy of the Investigators, though intended to answer questions that concerned me personally at the time I wrote it, is but little affected by relativity to that occasion. And I would make the same claim for my earlier piece on one of Coleridge's letters, and for my essay on Peirce at The Johns Hopkins.(20)

P. I take it this is the same question as that of subjectivity and objectivity on which we touched before.

H. Yes, and the obvious remark is that if there is a subjectivity or relativity which affects all history alike, it concerns the philosopher and not the historian. The only subjectivity that concerns the historian is that which is variable, which can be criticized, and which is corrigible. It is a subjectivity which is properly blamed, and is the opposite of an objectivity which is properly praised.

P. "Objective" means "value-free" here?

H. By no means. The historian is not blamed for praising and blaming, and praised for doing neither, but blamed if antecedent judgments of value blind him to contrary evidence, and praised if his selection and treatment of evidence is clearly not unbalanced by the desire to support judgments formed in advance of the search for evidence. As I intimated previously, the historian of art is a critic of art, the historian of religion a critic of religion, the historian of science a critic of science, and similarly the historian of economic, social, and political institutions is a critic of those institutions. Objectivity is not absence of criticism, but unreserved submission to further criticism, complete openness, withholding nothing from judgment. Furthermore, it is guaranteed not by the psychology or logic of the individual historian, but by the institutions of historical scholarship themselves, by the orderly procedures of review in the journals, by the ethics of bibliographic citation, by indexes, catalogues, bibliographies, reference works of all sorts, which protect the historian against involuntarily ignoring anything relevant to his undertakings, and provide ready means for calling him to account when he does.

P. That is, objectivity is institutionalized, and subjectivity is a weakness of the historian who is not yet institution-broken?

H. The analogy is far-fetched but, since you have fetched it, I won't ask you to throw it out if you will agree not to press it farther but to leave the historian otherwise unbroken.

P. Agreed. And now we have accounted for most of your published historical writing, and I still think that, in spite of your stubborn resistance, we have uncovered some covering laws, both laws that explain what the historians try to explain and laws that explain their explanations. But that will take some closer and more leisurely looking, and you may persuade me in the end that they are all nugatory. What bothers me now is this: If, as you say, the historian is, among other things, a critic of institutions, what is there left for philosophy to do? As no one knows better than you, the question is more embarrassing to me than it would be to many another philosopher, or than it would have been to me until a year ago, when, in a presidential address, I proposed to define philosophy itself as "the critic of institutions."(21)

H. I did not mean to needle you, but just to pull your leg gently. The historian criticizes only those institutions that come within the scope of his investigations. He does not often make explicit the principles of criticism upon which he relies. Still less often does he justify those principles as against alternatives that have been proposed by philosophers or assumed by other historians. And even when he does do so, his justifications are piecemeal and *ad hoc*, not systematic or comprehensive. So the philosophers need not fear that the historians will leave them little or nothing to do. But I seem to discern a continuum of inquiry connecting the shortest-tethered grubbing of the historian with the most abstract and universal critiques and speculations of the philosopher.

P. In any case, the ultimate problem of the philosophy of history may well be that of the relation between history and philosophy. Our symposium will not get that far. For the solution we had best look to philosophers who are also historians, or to philosophers who have carefully considered the work of such philosopher-historians. Meanwhile, you've given me enough for Donagan and Hinshaw to shoot at.

NOTES

1. This dialogue was read on May 3, 1957, as the opening paper of a symposium on the philosophy of history at the fifty-fifth annual meeting of the Western Division of the American Philosophical Association, at the University of Chicago. The papers of the two other participants have already been published: Virgil Hinshaw, Jr., "The Objectivity of History," *Philosophy of Science* 25: 51-58, 1958; Alan Donagan, "Social Science and Historical Antinomianism," *Revue internationale de philosophie* 11: 433-449, 1957.

2. Karl R. Popper: *Logik der Forschung*, 1935, 26 f. (English translation: *The Logic of Scientific Discovery*, 1959, 59 f.). Carl G. Hempel: "The Function of General Laws in History," *Journal of Philosophy* 39: 35-48, 1942, reprinted in: *Readings in Philosophical Analysis* selected and edited by Herbert Feigl and Wilfrid Sellars, 1949, 459-471. Karl R. Popper: *The Open Society and its Enemies*, revised edition, Princeton, 1950, 445-449, 720-723. Patrick Gardiner: *The Nature of Historical Explanation*, Oxford, 1952. (William Dray: *Laws and Explanation in History*, Oxford, 1957, appeared too late to be consulted.)

3. In Feigl and Sellars (cited in note 2), 465.

4. R. G. Collingwood: *The Idea of History*, Oxford, 1946, 249-282.

5. M.H.F.: *Stoicism and Roman Law*. Ph.D. thesis, typewritten, Cornell University Library.

6. Alan Donagan: "Explanation in History," *Mind* 66: 145-164, 1957.

7. M.H.F.: "Vico on Roman Law," in: *Essays in Political Theory presented to George H. Sabine* (edited by Konvitz and Murphy, Cornell University Press, 1948), 62-88. *The Autobiography of Giambattista Vico*, translated by M.H.F. and T.G. Bergin (Cornell University Press, 1944), 104 ff. *The New Science of Giambattista Vico*, translated by T.G.B. and M.H.F. (Cornell University Press, 1948), 347-353. Ferdinand Lassalle: *Nachgelassene Briefe und Schriften*, edited by G. Mayer, III, 387 f.

8. W.W.Tarn: "Alexander the Great and the Unity of Mankind," *British Academy Proceedings* 19: 123-166, 1933.

9. M.H.F.: "Alexander and the Stoics," *American Journal of Philology* 58: 59-82, 129-151, 1937.

10. Paul Russell Anderson and M.H.F.: *Philosophy in America from the Puritans to James*, 1939, 445 f.

11. M.H.F.: "Justice Holmes, the Prediction Theory of Law, and Pragmatism," *Journal of Philosophy* 39: 85-97, 1942; "Evolution in American Philosophy," *Philosophical Review* 56: 357-373, 1947, at 366f.

12. M.H.F.: "Alexander Bain and the Genealogy of Pragmatism," *Journal of the History of Ideas* 15: 413-444, 1954.

13. M.H.F. (editor): *Classic American Philosophers*, 1951, 1-39.

14. M.H.F.: "Dewey's Place in the Classic Period of American Philosophy," in: *Essays for John Dewey's Ninetieth Birthday* (edited by K.D. Benne and W.O. Stanley, Urbana, Illinois, 1950), 9-36.

15. M.H.F.: "The Academy of the Investigators," in: *Science, Medicine, and History: Essays on the Evolution of Scientific Thought and Medical Practice written in honour of Charles Singer* (2 vols., Oxford, 1953), I, 521-563.

16. M.H.F.: "Vesalius and His Book," *Bulletin of the Medical Library Association* 31: 208-221, 1943; "The Printer of Vesalius's *Fabrica*," *ibid.*, 240-259.

17. M.H.F.: "Vesalius in English State Papers," *Bulletin of the Medical Library Association* 33: 231-253, 1945.

18. M.H.F.: *Nicolaus Pol Doctor 1494.* New York: Herbert Reichner, 1947.

19. M.H.F. (editor): *Selected Papers of Robert C. Binkley* (Harvard University Press, 1948), 22.

20. See note 15 above. M.H.F.: "The Coleridges, Dr. Prati, and Vico," *Modern Philology* 41: 111-122, 1943; "Peirce at The Johns Hopkins University," in: *Studies in the Philosophy of Charles Sanders Peirce* (edited by Philip P. Wiener and Frederic H. Young, Harvard University Press, 1952), 277-311, 355-360, 363-374.

21. M.H.F.: "The Critic of Institutions," *Proceedings and Addresses of The American Philosophical Association* 29: 42-56, 1956.

Max Harold Fisch
A Bio-Bibliographical Sketch

by William B. Fisch

Max Harold Fisch was born in Elma, Washington, on December 21, 1900, the eldest son of a lumberjack in the coastal timber country. In 1906 the family moved to San Francisco, where the father became ship's barber on a trans-Pacific passenger steamship line, and the son attended the San Francisco public schools. At James Russell Lowell High School, where he graduated in January, 1919, Professor Fisch acquired a knowledge of Latin and Greek, the latter a highly unusual offering for a public school, both valuable assets for a future scholar. At the Panama-Pacific International Exposition in San Francisco in 1915, he competed successfully for admission to a demonstration school in the Palace of Education, where he received six months of intensive training in shorthand, typing, bookkeeping, and other commercial subjects. From 1916 to 1920 he taught these subjects in the University of California extension division.

By the time of his graduation from high school, Professor Fisch was actively involved in religion. At the insistence of a friend and former Sunday school teacher, he travelled to Indianapolis in 1920, took a job as secretary to the Executive Secretary of the Board of Education of the Disciples of Christ, and matriculated in Butler University, concentrating in philosophy. The bulk of his work in philosophy at Butler was done under Elijah Jordan, who was to remain the most influential of his teachers. While a student he served in rural churches as a lay minister for the Disciples of Christ, and was ordained in 1924. Following graduation in 1924 at the head of his class, Professor Fisch entered graduate school at Jordan's alma mater, Cornell University's Sage School of Philosophy.

Professor Fisch had hoped to study under James Creighton, then among the most influential creative American philosophers. Creighton's death early in the fall of 1924 frustrated this purpose; but he stayed on at the Sage School to complete his graduate study there, receiving his Ph.D. in 1930. He concentrated in ancient philosophy, the history of philosophy, and the philosophy of history, and wrote his dissertation under Dean William A. Hammond on the influence of Stoicism on Roman Law. From 1926 to 1928 he taught ethics and introduction to philosophy as an instructor in the Philosophy Department and served on the editorial staff of the Sage School's periodical, the *Philosophical Review.* In 1927 he married Ruth Bales, a graduate student in classics whom he had met while both were students at Butler.

In 1928 he joined the Philosophy Department in Western Reserve University at Cleveland, Ohio, where he remained until 1943. He taught a variety of courses, including ancient, Renaissance, and American philosophy, logic, and the history of philosophy. In ancient philosophy he published a series of articles on "Alexander and the Stoics" in 1937; in American philosophy he edited an anthology entitled *Philosophy in America from the Puritans to James* in collaboration with Paul R. Anderson, and published a seminal article on "Justice Holmes and the Prediction Theory of Law" in 1942. While at Western Reserve he became interested in the 18th Century Italian philosopher of history, Giambattista Vico. To further his study of Vico, Professor Fisch took a leave of absence in 1939 for research in Italy, where at the suggestion of the Vico scholar and biographer Fausto Nicolini he began work on translations of Vico's two greatest works, the *New Science* and the *Autobiography*, in collaboration with his friend Thomas G. Bergin. The Fisch-Bergin translation of the *Autobiography*, to which Prof. Fisch contributed an extensive interpretive introduction and notes, appeared in 1944, followed in 1948 by that of the *New Science*. These translations, since revised, are the only complete translations in English and have made possible a greatly expanded and deepened awareness of Vico among American students of philosophy and political science.

After the outbreak of the Second World War, Professor Fisch took the position of curator of the rare book collection of the Army Medical Library, which had been moved from Washington to Cleveland for safekeeping. Here he took up another scholarly theme for which he was prepared by the study of classical languages and literature—the history of science and medicine. He was curator of rare books from 1942-46, Chief of the History of Medicine Division in 1946, and co-editor of the Bulletin of the Medical Library Association from 1943-45. In this period he published a series of articles on Vesalius, a series on Nicholas Pol, a 16th Century court physician and book collector who promoted a highly popular and harmless but ineffective treatment for syphilis, as well as nearly a score of book reviews on various related subjects. He also served as an editor for *Blakiston's New Gould Medical Dictionary* (1949).

In September, 1946, Professor Fisch joined the faculty of the University of Illinois as professor of philosophy, a position he held until the retirement which is the occasion for this volume. As at Western Reserve, he taught a variety of courses and seminars at the undergraduate and graduate levels, but primarily in the history of ancient, medieval, and American philosophy. His scholarly work in American philosophy since 1946 has concentrated in what he calls the classic period, from the Civil War to the Second World War; his publications in this field include articles on "Evolution in American Philosophy" (1950), "Alexander Bain and the Genealogy of Pragmatism" (1954), and a summing-up of relationships, attitudes, and themes entitled "The Classic Period of American Philosophy", published as a general introduction to an anthology, *Classic American Philosophers* (1951). In the course of this historical study of American philosophy, Professor Fisch began collecting materials for a personal and intellectual biography of one of the most original thinkers of the period, Charles S. Peirce. In 1959 he was asked by Harvard University, custodian of the bulk of Peirce's unpublished manuscripts, to undertake an official biography; the completion of this monumental task, already crystallizing in the form of numerous articles and notes, has occupied a major portion of his time since then.

In the latter part of this classic period, Professor Fisch's attention has also been directed to the social philosophy of his college teacher, Elijah Jordan, whose work has found an increasing audience since his death in 1954. Fisch was a co-founder in the 1950's of the Jordan Society, a group of philosophers and social scientists, including others of Jordan's students, devoted to intensive study and development of his ideas; he edited and introduced an unfinished essay on Metaphysics by Jordan, published in 1956; and he contributed a preface to a study of Jordan's philosophy by George Barnett and Jack Otis in 1961. The influence of Jordan's experimental approach to the development, functioning, and evaluation of social institutions can be seen in Professor Fisch's own address, "The Critic of Institutions", delivered to the Western Division of the American Philosophical Association in 1955.

Interest in Vico and in the great 20th Century Italian social philosophers Benedetto Croce and Giovanni Gentile combined with interest in the history of science and medicine to motivate a year of study in Italy in 1950-51. The major portion of the year was spent in Naples, Vico's home, where in the course of research on Vico and his time, Professor Fisch studied a previously little-known school of the 18th Century Neapolitan scientists and physicians doing experimental work of considerable originality and significance. This group was the subject of an essay entitled "The Academy of the Investigators", published in 1953 in a *festschrift* for the historian Charles Singer. His work on Vico has included revisions of the introductions and translations of the *New Science* (1961, 1968, 1970) and the *Autobiography* (1963); a study of Vico's influence on Coleridge (1942); an article on his earlier work *Il Diritto Universale*, "Vico on Roman Law" (1948); and the article on Vico in the 1959 edition of *Encyclopaedia Britannica*. Professor Fisch's contributions to the history of science and medicine have also included numerous book reviews and a foreword to George Sarton's *The Life of Science*, (1948).

Recognition of Professor Fisch's stature as a philosopher and scholar has come in a number of ways. While in Italy in 1950, he was asked to represent the University of Illinois in the founding conference of the International Association of Universities, and he became a member of the organization's administrative board, serving in that capacity from 1950 to 1955. In 1954 he was elected vice-president and in 1955 president of the Western Division of the American Philosophical Association, and served as chairman of the national Board of Officers from 1956 to 1958. He was visiting professor of philosophy at the University of Chicago in the winter of 1955, and Matchette Lecturer at Purdue University in 1956. In 1958, with the support of a Fulbright grant, he accepted a longstanding invitation from a friend and fellow Cornell graduate student, now head of the English Department at Keio University, Tokyo, Japan, to come as visiting professor of American Philosophy. In February and March, 1958, prior to the beginning of the Japanese academic year, he lectured at many Indian universities on American philosophy under the auspices of the U.S. State Department. In the summer of 1965, he was visiting professor of philosophy at San Jose State University, California. Professor Fisch's distinguished teaching career, to the range and impact of which this volume seeks to testify, has not ended with his retirement. During the present academic year of 1969-70 he is visiting professor of philosophy at the State University of New York at Buffalo, teaching courses focusing on various aspects of the philosophy of Charles Peirce.

*University of North Dakota,
Grand Forks, North Dakota.*

A Bibliography of the Writings of Max Harold Fisch

by Willard M. Miller

(This bibliography of published material is intended to be complete through July 1, 1969. H. William Davenport and David E. Pfeifer have assisted me in its preparation.)

Abbreviations: BMLA, Bulletin of the Medical Library Association; JHMAS, Journal of the History of Medicine and Allied Sciences; JP, Journal of Philosophy; PR, Philosophical Review.

1927

Review of Wilhelm Jerusalem, *Einführung in die Soziologie*, and Walther Eckstein, *Das antike Naturrecht in sozialphilosophischer Beleuchtung*. *PR* 36: 592-593.

1928

Review of Aristotle, *The Nicomachean Ethics*, text with an English translation by H. Rackham (Loeb Classical Library). *PR* 37: 282-284.

Review of William P. Montague, *The Ways of Knowing: or, The Methods of Philosophy*. *American Journal of Psychology* 40: 322.

Review of G. Watts Cunningham, *Problems of Philosophy*, and Roy Wood Sellars, *The Principles and Problems of Philosophy*. *American Journal of Psychology* 40: 323-324.

Review of Marcus Tullius Cicero, *On the Commonwealth*, translated with notes and an introduction by George H. Sabine and Stanley B. Smith. *PR* 39: 622-625.

Stoicism and Roman Law. Ph.D. thesis. Cornell University. Typewritten. Cornell University Library.

1933

Review of Wilbur M. Urban, *Fundamentals of Ethics*. *PR* 42: 85-88.

1934

Review of Susanne K. Langer, *The Practice of Philosophy*. *PR* 43: 325.

1935

Review of *Studies in the Nature of Facts* and *Causality* (University of California Publications in Philosophy, Vol. XIV and Vol. XV). *PR* 44: 590-593.

1936

Review of *Spinoza nel terzo centenario della sua nascita* (supplemento speciale al vol. XXV della *Revista di filosofia neo-scolastica*) and Fausto Meli, *Spinoza e due antecedenti italiani dello spinozismo*. *PR* 45: 326-327.

1937

Review of Lawrence J. Henderson, *Pareto's General Sociology*. *PR* 46: 438-439.

"Alexander and the Stoics," *American Journal of Philosophy* 48: 59-82, 129-151.

1939

Philosophy in America from the Puritans to James, edited in collaboration with Paul Russell Anderson. New York: D. Appleton-Century Co.

1940

Review of A. J. Festugière, *Contemplation et vie contemplative selon Platon*. *PR* 49: 77-80.

Review of Jacques Barzun, *Of Human Freedom*. *Mississippi Valley Historical Review* 27: 513-514.

1942

"Justice Holmes, the Prediction Theory of Law and Pragmatism," *JP* 39: 85-97.

1943

"Kos and Salerno," *Bulletin of the Academy of Medicine of Cleveland* 28: no. 11, 10-11.

"Joint Authorship," *Current List of Medical Literature* 5: nos. 13-14, inside front cover.

"Vesalius and His Book," *BMLA* 31: 208-221.

"The Printer of Vesalius's *Fabrica*," *BMLA* 31: 240-259.

"The Coleridges, Dr. Prati, and Vico," *Modern Philology* 41: 111-122.

Review of Stephen Miswa, *Nicolas Copernicus*. *BMLA* 31: 283.

Review of Howard B. Adelmann, *The Embryological Treatises of Hieronymus Fabricius of Aquapendente*. *BMLA* 31: 376-379.

Review of Gregory Zilboorg, *Mind, Medicine, and Man*. *BMLA* 31: 386-387.

1944

The Autobiography of Giambattista Vico, translated from the Italian in collaboration with Thomas Goddard Bergin with an introduction, notes and a chronological table by Max H. Fisch. Ithaca: Cornell University Press. Second printing (revised), November, 1944.

Review of Harvey Cushing, *A Bio-Bibliography of Andreas Vesalius*. *BMLA* 32: 113-115.

Review of James F. Ballard, *A Catalogue of the Medieval and Renaissance Manuscripts and Incunabula in the Boston Medical Library*. *BMLA* 32: 398-401.

"Rare Book Collection of Army Medical Library," *Bulletin of the Academy of Medicine of Cleveland* 29: no. 11, 9.

1945

"Vesalius in English State Papers," *BMLA* 33: 231-253.

Review of Harry Friedenwald, *The Jews and Medicine*. *BMLA* 33: 113.

Review of Pascual Jordan, *Physics of the 20th Century*. *BMLA* 33: 116.

Review of Benjamin W. McCready, *On the Influence of Trades*, edited with an introduction by Genevieve Miller. *BMLA* 33: 116.

Review of Aristides A. Moll, *Aesculapius in Latin America. BMLA* 33: 117-118.

Review of George Rosen, *The History of Miners' Diseases. BMLA* 33: 120-121.

Review of Laurence Stapleton, *Justice and World Society. PR* 54: 277-280.

Review of Bergen Evans, *The Psychiatry of Robert Burton. BMLA* 33: 370-372.

Review of Bernard Jaffe, *Men of Science in America. BMLA* 33: 378-379.

Review of *Pharmacopoeia Londinensis of 1618*, facsimile edition with a historical introduction by George Urdang. *BMLA* 33: 541-542.

Review of Owsei Temkin, *The Falling Sickness: A History of Epilepsy from the Greeks to the Beginnings of Modern Neurology. BMLA* 33: 546-547.

Review of J.M.D. Olmstead, *François Magendie, pioneer in experimental physiology and scientific medicine in xix century France. BMLA* 33: 547-550.

Review of Sidney Warren, *American Freethought, 1860-1914. PR* 54: 614-615.

1946

"Nicholaus Pol Doctor 1494," *Bulletin of the History of Medicine* 20: 294-298.

"Nicholaus Pol and Veit Bild," *JHMAS* 1: 677-681.

Review of Henry Alonzo Myers, *Are Men Equal? : An inquiry into the meaning of American democracy. PR* 55: 101-103.

Review of Emma J. and Ludwig Edelstein, *Asclepius: A Collection and Interpretation of the Testimonies. Bulletin of the History of Medicine* 20: 712-716.

"Notes & Queries," *JMHAS*

1947

Nicolaus Pol Doctor 1494, with a critical text of Pol's Guaiac Tract edited and translated by Dorothy M. Schullian. New York: Herbert Reichner.

"Evolution in American Philosophy," *PR* 56: 357-373.

"The Marshall Collection of Herbals," in collaboration with Ruth B. Fisch, *Bulletin of the History of Medicine* 21: 224-261.

"The Mather Bibliographies," discussion of Thomas J. Holmes's bibliographies of Increase and Cotton Mather, *JHMAS* 2: 121-123.

Review of Herbert W. Schneider, *A History of American Philosophy. Journal of the History of Ideas* 8: 484-491.

"Notes & Queries," *JHMAS* 2: 121-128, 262-267.

1948

The New Science of Giambattista Vico, translated from the Italian in collaboration with Thomas Goddard Bergin. Ithaca, New York: Cornell University Press.

Selected Papers of Robert C. Binkley, edited with a biographical sketch and a bibliography; foreword by Luther Evans. Cambridge, Mass.: Harvard University Press.

"Vico on Roman Law," *Essays in Political Theory*, presented to George H. Sabine, edited by Milton R. Konvitz and Arthur E. Murphy. Ithaca, New York: Cornell University Press, 62-68.

Preface in *A Catalogue of Incunabula and Manuscripts in the Army Medical Library*, by Dorothy Schullian and Francis E. Sommer. New York:

Henry Schuman, vii-x.

Foreword in *The Life of Science: Essays in the History of Civilization*, by George Sarton. New York: Henry Schuman, v-vii.

Review of Herbert W. Schneider, *A History of American Philosophy*, and Joseph L. Blau, *American Philosophical Addresses 1700-1900*. *American Literature* 19: 380-382.

Review of James B. Conant, *On Understanding Science: An Historical Approach*. *JHMAS* 3: 446-449.

Review of Kurt F. Leidecker, *Yankee Teacher: The Life of William Torrey Harris*. *New England Quarterly* 21: 117-119.

1949

(Member of Editorial Board and contributor to:) *Blakiston's New Gould Medical Dictionary*. Philadelphia: The Blakiston Co., 1949.

Review of Mario M. Rossi, *La Vita, le opere, i tempi de Edoardo Herbert di Chirbury*. *JP* 46: 195-203.

Review of *Manuscripts of Leonardo da Vinci*, Elmer Belt Library of Vinciana. *JHMAS* 4: 243-244.

Review of *Atti del Congresso internazionale di filosofia promosso dall' Istituto di studi filosofici, Roma 15-20 novembre 1946*, Vol. I, Il materialismo storico, a cura di Enrico Castelli. *PR* 58: 278-281.

Review of Benedetto Croce, *Bibliografia Vichiana*, Giambattista Vico, *Autobiografia*, a cura di Fausto Nicolini, and Fausto Nicolini, *Uomini di spada di chiesa di toga di studio ai tempi di Giambattista Vico*. *PR* 58: 528-529.

Review of Vernon L. Parrington, Jr., *American Dreams: A Study of American Utopias*. *Journal of English and Germanic Philology* 48: 166.

1950

"Dewey's Place in the Classic Period of American Philosophy," *Essays for John Dewey's Ninetieth Birthday*, edited by Kenneth D. Benne and William O. Stanley. Urbana: University of Illinois Press, 9-36.

Review of W. H. Werkmeister, *A History of Philosophical Ideas in America*, and Philip P. Wiener, *Evolution and the Founders of Pragmatism*. *American Literature* 22: 185-189.

1951

Classic American Philosophers, general editor and author of general introduction. New York: Appleton-Century-Crofts, Inc.

Review of *The Renaissance Philosophy of Man*, edited by Ernst Cassirer, P. O. Kristeller, and J. H. Randall, Jr. *PR* 60: 109-111.

1952

"Peirce at the Johns Hopkins University," with Jackson I. Cope, *Studies in the Philosophy of Charles Sanders Peirce*, edited by Philip P. Wiener and Frederic H. Young. Cambridge: Harvard University Press, 277-311, 355-360, 363-374.

"Some Additions to Morris R. Cohen's Bibliography of Peirce's Published Writings," with Daniel C. Haskell, *ibid.*, 375-381.

Review of Caelius Aurelianus, *On Acute Diseases and On Chronic Diseases*, edited and translated by I. E. Drabkin. *Classical Weekly* 45: 157.

Review of Joseph L. Blau, *Men and Movements in American Philosophy*. *JP* 49: 675-677.

"Prefacio" in *La Vida de la Ciencia: Ensayos de Historia de la Civilización* por George Sarton. Buenos Aires: Espasa-Calpe Argentina.

1953

"The Academy of the Investigators," *Science, Medicine, and History: Essays on the Evolution of Scientific Thought and Medical Practice*, written in honor of Charles Singer. 2 vols. Oxford: At the University Press, 1: 521-563.

1954

"The International Association of Universities," *The Educational Record* 35: 157-159.

"Alexander Bain and the Genealogy of Pragmatism," *Journal of the History of Ideas* 15: 413-444.

Review of J. H. G. Grattan and Charles Singer, *Anglo-Saxon Magic and Medicine. JHMAS* 9: 252-253.

Review of George Sarton, *A History of Science: Ancient Science through the Golden Age of Greece. Bulletin of the History of Medicine* 28: 280-287.

Review of Richard D. Mosier, *The American Temper: Patterns of Our Intellectual Heritage. JP* 51: 445-446.

1955

"Elijah Jordan," memorial note, *Proceedings and Addresses of the American Philosophical Association* 28: 62-63.

"Benedetto Croce," *Twentieth Century Encyclopedia of Religious Knowledge* 1: 312-313.

Review of Ewart Lewis, *Medieval Political Ideas. Ethics* 66: 75.

1956

"The Critic of Institutions," *Proceedings and Addresses of the American Philosophical Association* 29: 42-56.

Metaphysics, an unfinished essay by Elijah Jordan, edited with a preface and bibliography. Evanston: Principia Press of Illinois.

Review of A. Robert Caponigri, *History and Liberty: The Historical Writings of Benedetto Croce. New Scholasticism* 30: 379-380.

1957

Review of A. Robert Caponigri, *Time and Idea: The Theory of History in Giambattista Vico. JP* 54: 648-652.

1959

"Giambattista Vico," *Encyclopaedia Britannica.* 23: 122-123.

"Keio 2058: The Nature and Function of a University" (in Japanese translation, "Keio 2058 nen: Daigaku no Hinshitsu to Sekinin"), *Keio Shinbun* number 241, January 15, p. 6. (Revised translation, "Daigaku wa ikaniarubekika: Keiogijuku no kitarubeki Hyankuen no tameni"), *Mita Hyoron* (Mita Review), number 583, July, pp. 2-5.

"The University Ideal," *Bulletin of the International Association of Universities* 7: 125-127.

1960

"The Philosophy of History: A Dialogue," *Philosophy* (Tetsugaku, Mita Philosophy Society, Keio University, Tokyo) 36: 149-170.

"The Idea of Institution in the Major Religions," *Proceedings of the Contemporary Philosophy*, edited by Seizi Uyeda. Tokyo: Waseda University Press, 473-477.

1961

The New Science of Giambattista Vico, in collaboration with Thomas Goddard Bergin, abridged, revised, and with a new introduction. Garden City, New York: Doubleday and Company, Inc.

Foreword in *Corporate Society and Education: The Philosophy of Elijah Jordan*, by George Barnett and Jack Otis. Ann Arbor: University of Michigan Press, v-vi.

1962

Review of Edward C. Moore, *American Prgamatism: Peirce, James, and Dewey. American Literature* 34: 138-140.

1963

The Autobiography of Giambattista Vico, in collaboration with Thomas Goddard Bergin, revised with supplementary notes. Great Seal Books edition. Ithaca, New York: Cornell University Press.

1964

"A Chronicle of Pragmaticism, 1865-1879," *Monist* 48: 441-466.

"The American Proposition," *Pacific Philosophy Forum* 2: no. 3, 96-100.

"Was There a Metaphysical Club in Cambridge?" *Studies in the Philosophy of Charles Sanders Peirce, Second Series*, edited by Edward C. Moore and Richard S. Robin. Amherst: University of Massachusetts Press, 3-32.

"A First Supplement to Arthur W. Burks's Bibliography of the Works of Charles Sanders Peirce," *ibid.*, 477-485.

"A Draft of a Bibliography of Writings about Charles Sanders Peirce," with the assistance of Barbara E. Kretamann and Victor F. Lenzen, *ibid.*, 486-514.

Review of John E. Smith, *The Spirit of American Philosophy. American Literature* 36: 239-241.

Review of Paul R. Anderson, *Platonism in the Midwest. History of Education Quarterly* 4: 311-312.

"Philosophical Clubs in Cambridge and Boston from Peirce's Metaphysical Club to Harris's Hegel Club," *Coranto* 2: no. 1, 12-23. This is the first paper in a series of three under this title.

1965

"Philosophical Clubs in Cambridge and Boston from Peirce's Metaphysical Club to Harris's Hegel Club," *Coranto* 2: no. 2, 12-25; 3: no. 1, 16-29. These are the second and third papers in the series of three under this title.

Preface in *European Philosophy Today*, edited by George L. Kline. Chicago: Quadrangle Books, 5-9.

1966

Classic American Philosophers, general editor of this revised edition with a new preface. New York: Appleton-Century-Crofts, Inc.

"A Second Supplement to Arthur W. Burks's Bibliography of the Works of Charles Sanders Peirce," *Transactions of the Charles S. Peirce Society* 2: 51-53.

"A First Supplement to 'A Draft of a Bibliography of Writings about Charles Sanders Peirce'," *ibid.*, 2: 54-59.

"Peirce's Triadic Logic," with Atwell R. Turquette, *ibid.*, 2: 71-85.

"Chicks Under Glass," review of Howard B. Adelmann, *Marcello Malpighi and the Evolution of Embryology. New York Times Book Review*, September 18, 7, 47.

1967

"Peirce's Progress from Nominalism Toward Realism," *Monist* 51: 159-178.

"Marcello Malpighi e l'evoluzione dell' embriologia" (Italian translation of "Chicks Under Glass"), broadcast over Radio-televisione Italiana March 9 and distributed by United States Information Service as Testo n. 823, La Voce dell' America.

1968

The New Science of Giambattista Vico, revised translation in collaboration with Thomas Goddard Bergin, with new preface and introduction. Ithaca: Cornell University Press.

"Croce e Vico," *Rivista di Studi Crociani* 5: 5-30, 151-171 (translated and edited by Maria Donzelli).

1969

"Vico and Pragmatism," *Giambattista Vico: An International Symposium,* edited by Giorgio Tagliacozzo and Hayden V. White. Baltimore: Johns Hopkins Press, 401-424.

University of Vermont,
Burlington, Vermont.

THESES DIRECTED BY MAX H. FISCH AT THE UNIVERSITY OF ILLINOIS 1946–1969

1947. Easterling, Marvin Leon. Santayana's Theory of Knowledge. M.A.

1949. Fain, Haskell. A Comparison of the Theories of Logic of John Dewey and Charles Sanders Peirce. M.A.

——. Haynes, Thomas Morris. Institutional Theories of Law: Hauriou and Jordan. Ph.D.

1950. Haworth, L. L. Political Freedom: Hegel and John Stuart Mill. M.A.

——. Skeen, Nancy Noble. Psychology as a Basis for Ethics. M.A.

1952. Haworth, L. L. The Practical Philosophies of John Dewey and Elijah Jordan. Ph.D.

1953. Cornett, Robert Arnold. Individualism in the Ethics of Elijah Jordan. Ph.D. ——. Gardner, James Russell. The Role of Experience and Value in Naturalistic and Personalistic Thought as Represented by the Philosophies of Clarence Irving Lewis and Edgar Sheffield Brightman. Ph.D.

——. Lunine, Myron Julius. An Examination of Charles Peirce's Metaphysical Generalization of Charles Darwin's Scientific Theory of Evolution. M.A.

——. Russell, Elbert Winslow. Elijah Jordan's Theory of Obligation as Stated in *The Good Life*. M.A.

1954. Harris, Henry Silton. The Social Philosophy of Giovanni Gentile. Ph.D.

——. Kimura, Douglas Shuso. Reason and the Passions in Hume's Moral Theory. M.A.

1955. Bosworth, John Richard. The Concept of Feeling in the Philosophy of Elijah Jordan. M.A.

——. Howe, Lynette Sue. George Santayana and the Pragmatists. M.A.

1956. Celarier, James LeRoy. Reason and Desire in the Ethics of Aristotle. B.A.

——. Hardeman, Thomas Patterson. The Philosophy of Lucius Annaeus Seneca. Ph.D.

——. Jovan, George Anthony. Theory of Perception in the Later Works of Alfred North Whitehead. B.A.

1957. Sheridan, James Francis. Paul Carus: A Study of the Thought and Work of the Editor of the Open Court Publishing Company. Ph.D.

1958. Celarier, James LeRoy. Plotinus's Treatment of the Categories. M.A.

——. Haynes, Richard Pierce. The Material Cause as an Object of Knowledge in Aristotle. M.A.

——. Stannard, Jerry Willmert. The Psychology of the Passions in the Old Stoa. Ph.D.

——. Suits, Bernard Herbert. The Aesthetic Object in Santayana and Dewey. Ph.D.

1960. Carloye, Jack Cloyd. Reason as a Natural Function in the Philosophy of John Dewey. (Co- director: Harry M. Tiebout.) Ph.D.

1961. Ficarra, Francis Thomas. Collingwood's *New Leviathan*. Ph.D.

1962. Haynes, Richard Pierce. Plato's Theory of Forms and the Self-Predication Assumption. Ph.D.

1963. Min, Yung Wha Kim. Pragmatism and Realism in Peirce's Philosophy. M.A.

——. Minas, James Sayer. On the Construction of Propositional Calculi. Ph.D.

——. Roberts, Don Davis. The Existential Graphs of Charles S. Peirce. Ph.D.

1964. Fan, Kuang Tih. Charles S. Peirce's Theory of Abduction. M.A.

——. Tursman, Richard Allen. Nietzsche's View of Science and Art. Ph.D.

1966. Grossmann, Ross Milton. Abduction, Simplicity and Instinct in the Philosophy of C. S. Peirce. M.A.

——. Koehn, Donald Robert. Charles S. Peirce's "Illustrations of the Logic of Science". M.A.

——. Miller, Richard Warner. Charles S. Peirce's First Series of Papers in Logic. M.A.

——. Miller, Willard Marshall. The Synechistic Philosophy of Charles Sanders Peirce: A Hypothetical Cosmology in Accord with the Logic of Science. B.A.

——. Pfeifer, David Elmer. Some Difficulties Concerning Aristotle's Doctrine of Substance in the *Categories*. M.A.

——. Swanson, Gerald William. The Intellectual Development of Chauncey Wright. M.A.

1968. Douglas, George Halsey. Croce's Aesthetic and Theory of Culture in *The Philosophy of Spirit*. Ph.D.

——. Miller, Willard Marshall. History as Science in the Philosophy of Charles Sanders Peirce. M.A.

1969. Brock, Jarrett Ernest. C. S. Peirce's Logic of Vagueness. Ph.D.

——. Cashdollar, Stanford E., Jr. An Inquiry into Aristotle's Ethics and Politics: Is There an Autonomous Science of Morals? Ph.D.

——. Davenport, Herbert W. The Career of Boyd H. Bode in the University of Illinois Department of Philosophy—1909-1921. M.A.

——. Fu, Wei-Hsun. Contemporary Ethical Autonomism—A Critical Study of Sartre and Hare. Ph.D.

——. Harter, Edward Darcy. Aristotle's Two Doctrines of Primary Substance: A Comparative Analysis. M.A.

——. Koehn, Donald Robert. Peirce's Explanation of the Validity of Synthetic Inference in the "Illustrations of the Logic of Science". Ph.D.

——. Miller, Richard Warner. Charles S. Peirce's Theory of Probability. Ph.D.

——. Miller, Willard Marshall. C. S. Peirce On the Philosophy of History. Ph.D.

——. Peterson, John Austin. Civil Disobedience. B.A.

——. Roth, Michael David. An Examination of Plato's *Cratylus*. Ph.D.

——. Thompson, John Howard. Francis Lieber's Theory of Institutions. M.A.

Nearing Completion as of August 31, 1969

Casper, Dennis John. Being and Predication in Plato's *Sophist.* Ph.D.

Pfeifer, David Elmer. Pragmaticism and the *Summum Bonum:* A Study in the Philosophy of C. S. Peirce. Ph.D.